# Rambling Through
## SCIENCE AND TECHNOLOGY
*Fourth Edition*

by
Octave Levenspiel

Octave Levenspiel
Chemical Engineering Department
Gleeson Hall
Oregon State University
Corvallis, Oregon, 97331-2702

E-mail   levenspo@peak.org
Phone   (541) 753-9248
Fax      (541) 752-1755

Fourth Edition published April 2013

© 2013 Octave Levenspiel
All Rights Reserved
ISBN 978-1-300-68720-7
Printed and Distributed by Lulu.com
Graphics by Bekki Levien, BLDesigns.biz

# Contents

|    | Preface |             |
|----|---------|-------------|
| 1  | The Birth of Science | 1.1 - 1.11 |
| 2  | Truth and Knowledge | 2.1 - 2.5 |
| 3  | Reason, Logic and Deduction | 3.1 - 3.12 |
| 4  | Mathematics | 4.1 - 4.16 |
| 5  | Inductive Reasoning | 5.1 - 5.4 |
| 6  | Probability | 6.1 - 6.12 |
| 7  | Statistics | 7.1 - 7.32 |
| 8  | Decision Theory | 8.1 - 8.7 |
| 9  | Theory of Games | 9.1 - 9.22 |
| 10 | Galileo and Newton, Kinematics and Dynamics | 10.1 - 10.11 |
| 11 | The Story of Science | 11.1 - 11.11 |
| 12 | The First Law and the Concept of Energy | 12.1 - 12.12 |
| 13 | The Amazing Second Law | 13.1 - 13.22 |
| 14 | Fields - Gravitational, Electric and Magnetic | 14.1 - 14.9 |
| 15 | Measures of This and That | 15.1 - 15.4 |
| 16 | Cosmology and the Universe | 16.1 - 16.18 |
| 17 | Thoughts About the Universe | 17.1 - 17.15 |
| 18 | The Last Word | 18.1 |
|    | Index | $i - vi$ |

# A Few Words of Explanation

As a long-time retiree I have a bookcase and a few file cabinets crammed with my notes and thoughts. I never was a throw-awayer. And since I've given up running the marathon I thought it a good idea to try to make some sort of order of those jumbles of notes.

I did this a few years ago and ended up with a three-ring binder of thermo notes which were turned into a book entitled "Understanding Engineering Thermo". And recently I came up with another three-ring binder of general notes. My daughter, Bekki Levien said "Let me try to make a book of this". I took her up on this.

This is the result - random notes on a variety of topics, from the beginnings of science to astronomers' ideas today of the universe. In this adventure we touch on the history of science, the heroes of science, the fantastic invention of the heat engine, how to tell which is the most efficient of mechanical devices, the recent creation or invention of the theory of statistics and also the theory of games (so important in resolving conflicts), plus various other topics including Einstein and Carnot and that mysterious non-thing called entropy. Enjoy.

I wish to thank my childhood friend Denis Kliene, who introduced me to the Laughing Dragon, who adopted me and became my personal Joss. And to 'can do' Bekki who pushed me to write this book and made most of the drawings plus the cover of this book, an enormous thanks to you, Bekki. And finally, Tom Fitzgerald and Ron Hershel who gave me arguments galore; thanks for keeping me on track.

—*Octave Levenspiel*

P.S. As always I welcome comments on what I have written.

Here's how to reach me:  address: 2135 NW Elmwood
Corvallis OR 97330

e-mail: levenspo@peak.org

phone: (541) 753-9248

CHAPTER **1**

# THE BIRTH OF SCIENCE

## THE ANCIENT WORLD AND RATIONALISM

At the beginnings of recorded time, man's world history was filled with conflicting beliefs, strange stories of benevolent and terrifying monsters and gods, and mystic rites. Such myths were used to explain creation, birth, death, love, hate and the other mysteries of life. Since the dawn of time peoples everywhere have created such imaginary worlds to help them through the difficulties of life. These myths lie at the intersection of imagination and history, dreams and reality.

China is among the oldest of the world's civilizations, and central in Chinese mythology is the dragon, a benevolent monster capable of doing great harm but generally portrayed as man's protector which keeps watch over rain, clouds, and winds. These beneficent powers were appreciated by China's rulers who adopted the dragon as the imperial symbol.

Other peoples: Hindus, Egyptians, Assyrians, Hebrews, Mayas, Aztecs, and innumerable others all had their elephant gods, their giant turtles that supported the earth, stories and myths about how things had started and how we could coax the powers to aid us in our lives, by prayer or by offering gifts. Most of these stories were created before writing was invented so they have come to us by word of mouth.

In Europe the earliest written record of peoples' thoughts and ideas have come to us from the writings of the Greeks about 600-500 BC. Their mythology was peopled with a whole world of gods, Zeus being the chief. But Zeus had many human frailties, including being rather sneaky and unfaithful to his wife.

Most of these Greek writings came to European scholars and then to us by a long circuitous route—first to the world's premier library, the Great

Library at Alexandria, then by Arabic scholars who took this information west across North Africa, then to Spain, thence to European scholars around year 1000 AD.

**Since today's science comes in large part from these Greek developments let us focus on them.**

These ancient ideas so dominated early man's thoughts that he felt that they controlled his actions and his future. This led him to believe that all reason and knowledge of the world came from his mind. For close to 2000 years, until about 1500 AD European scholars accepted this Greek concept of knowledge, that:

> *"knowledge, all knowledge, about gods and dragons and*
> *how the world works come from your mind,*
> *your reasoning and only your reasoning"*

This philosophical view is called **rationalism**.

Thus for 2000 years, until about the 16th century men touted a whole marketplace of ideas, some favoring this idea, others favoring that. No one idea represented the truth. You just accepted the idea that you thought to be most reasonable. It was a very democratic view of the world.

For example some believed that the world was flat and perfectly square (Chinese myth), a flat circular disc with Mt. Olympus at its center (Greek myth), a sphere spinning about the sun (Aristarchus' idea), a stationary sphere at the center of the universe with everything else rotating about it (Aristotle's idea). Of course, I feel that a sausage shaped Earth was the most reasonable.

Let me then ask - how many teeth has a lion? You may say 32 but I say 28. Why? Because 28 = 4 x 7 and seven is a very special number. Note that:
- the world was made in 7 days,
- there are 7 days in a week
- there are 7 wanderers in the sky - Sun, Moon, Mercury . . .
- there are 7 entries to your soul - eyes, ears, nostrils, mouth.

Here I have given you four good reasons for 28 teeth. What reason have you to claim 32 teeth? Please tell me.

In the physics of those times there was also Aristotle's basic law of mechanical action which tells why a thrown ball keeps moving when it has left your hand. Here is his law:

**"nature abhors a vacuum"**

So when you throw a ball a vacuum is created behind the moving ball, air rushes in and pushes the ball forward. It reminds one of the saying: "lifting oneself up by ones' bootstraps". Related to this was the idea that

**"a continually acting force is needed to maintain motion."**

Forces were considered to be of two types, natural forces and the unnatural forces. Another one of Aristotle's rules also said that only one kind of force was able to act on a body at any one time. These rules were used by Santbach in his 1561 treatise on gunnery. There he said that a projectile from a cannon travelled in a straight line until its velocity in that direction became zero. After that the force of gravity would be able to act upon it. This would then cause the cannonball to fall vertically to the ground, and of course the bigger the cannonball the faster it would fall (another one of Aristotle's laws). Santbach's book was the standard text for military gunners of armies aiming to pulverize enemy castle battlements.

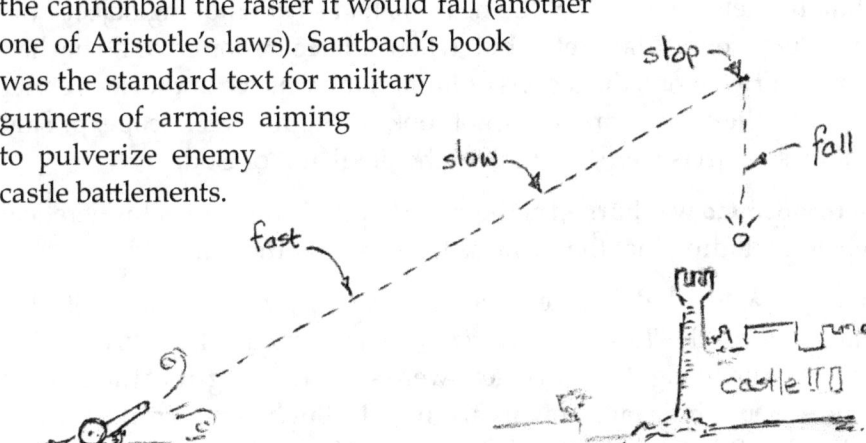

These are a few examples of rationalist thinking that dominated this 2000 year period. As Einstein wrote:

*"Rationalism is the theory that says that reason in itself is the source of all knowledge, superior and independent of sense perception"*

## The Adoption of Aristotle's Ideas by the Church and the Growth of Authoritarianism

Around 1200 AD saw an important event in the history of knowledge, and a split in rationalism. At that time Thomas Aquinas coaxed and convinced the Church to adopt Aristotle's ideas of truth and welded these ideas to Christian theology. This was a disastrous event in the prehistory of science since it stopped any questioning on these matters. From then on it was no longer safe to harbor contrary ideas, least of all that the earth was round and circled the sun. This represented an upheaval in rationalism, replacing the variety of ideas by an authoritarianism by the Church. Thus if you expressed an idea contrary to the Church's that was heresy, and for your own good you had to be purified of these false ideas. Here are some examples:

**Copernicus**, a priest, thought that the sun was the center of the universe, not the earth, but delayed for 30 years until his dying days to have his ideas published for fear of punishment from the church authorities.

Church teachings said that the heavens had seven moving objects: the Sun, our Moon, Mars, etc., but when **Galileo** reported that he had observed two more, the moons of Jupiter, the Church authorities either did not believe him, or would not look through his telescope, or said that these objects were the work of the devil, not of God.

**Giordano Bruno** was burned at the stake to purify his soul of his heretical beliefs, including that the sun was the center of the universe.

Finally, **Galileo**, Italy's chief scientist, had the courage to write his Dialogues on the *Two Chief World Systems*. He was old, infirm, going blind, but he thought that his ideas were so convincing that the Church fathers would be swayed by his arguments. But he was wrong. He was ordered to Rome to stand trial for heresy. Here is how it went:

> On June 22, 1633 this elderly man sank stiffly to his knees in the great hall of the convent of Santa Maria Sopra Minerva in Rome. Before him were the cardinals of the Inquisition. Through a door to one side, it is said he could see the torturer standing ready. He started reading from the document that had been handed to him:

> *"I, Galileo, . . . my age being seventy years, having been called personally to judgement and kneeling before your Eminences, Most Reverent Cardinals, General Inquisitors against heretical depravity . . . do swear that I have always believed, do now believe, and with God's aid shall believe hereafter all that is taught and preached by the Holy Catholic and Apostolic Church. I wholly reject the false opinion that the sun is the center of the world and moves not . . . I wrote and published a book in which the said condemned doctrine was treated and gave very effective reasons in favor of it without suggesting any solution. Because of this I am by this Holy Office judged vehemently of heresy . . ."*

Galileo was found guilty, served a mild sentence of life imprisonment (house arrest), all copies of his book in Church territories were collected and burned, and his book was put on the Index of Forbidden Books, which meant that if you read this book you went straight to hell. However some copies found their way to non Church territories where they made a strong impression and were widely read and copied.

This trial of Galileo, more than anything else, divided Europe into two regions. In the Church dominated regions the teachers and scholars accepted the teachings of Aristotle unquestioningly as the truth because they were said to be the words of God. There was no need to do research on those matters - on falling bodies, on where the center of feeling and thought was in your body (guess where they knew it was - in your heart? your head? your stomach?).

If a Martian would observe us today he would observe us signing love letters with hearts, he would see the heart symbol for Valentine's Day, and see us place our hands on our hearts when we salute the flag - not our stomach or our head. In the rest of Europe, England, Holland, Germany, and so on, the location of these feelings was still open to question.

Until the 1500's Italy led Europe in new ideas, great art, great architecture, etc. But the Church's harsh clampdown on new ideas, in particular the punishment of Galileo, cast a chill on further attempts to explore new ideas about the world around us. From then on Italy stagnated, while the rest of Europe took the lead in the quest for knowledge.

## The Birth of the Scientific Age

Around the 1500-1600's AD a thinking revolution occurred which changed our whole view of the world and brought on the scientific age. This revolution was due to a few mechanical inventions, plus the trial of Galileo. (We will go into this matter further in Chapter 10).

The **first** invention was the mechanical clock, which allowed man to measure time intervals reasonably well, instead of using the human pulse, the sundial, burning candles, and other such crude measuring devices.

**Second**, we have the invention of the thermometer. Instead of measuring the degree of hotness of an object qualitatively with a scale such as

Extreme Hot
Very Hot
Hot
Warm
Temperate
Cold
Frost
Hard Frost
Great Frost
Extreme Frost

numerical scales were invented. The best known was the Fahrenheit scale.

**The Fahrenheit Temperature Scale**

Everyone has roughly the same body temperature, so let us call it 100°.

Before the invention of refrigeration, no one was able to get a temperature below that of a mixture of water, ice and salt. So this was called 0°.

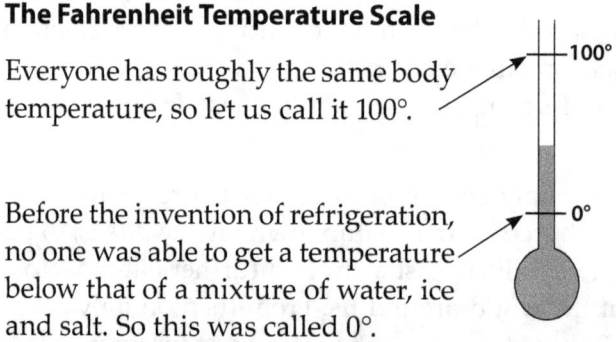

The **third** invention was the lens which allowed the user to manipulate images and make the spyglass, the magnifying lens, the telescope, reading glasses, etc.

> **The genius of Galileo was the final catalyst to bring on the scientific revolution.**

## THE BIRTH OF EMPIRICISM

With these tools available man could measure the physical world, and when he did he found that every one of Aristotle's truths about what we call physics today was wrong.  So a new idea gained credence for scholars, and that was:

**All knowledge about the world comes from measurements of the real world.**

No truths originate from your mind. Everything you say about the real world has to have a correspondence with the world. This approach to knowledge is called **Empiricism**.

## GALILEO'S EXAMPLE

In one of Galileo's dialogues on physics, the character representing the empiricist speaks to the rationalist and says

> *"Aristotle claims that "an iron ball of 100 pounds falling from a height of one hundred cubits reaches the ground before a one-pound ball has fallen a single cubit", I say that they arrive at the same time. You find, on making the experiment that the larger outstrips the smaller by two finger-breadths. Now you would not hide BEHIND these two finger-breadths the ninety-nine cubits of Aristotle, nor would you mention my small error and at the same time pass over in silence his very large one?"*

'The Two New Sciences' by Galileo Galilei. Translated from the Italian and Latin by H. Crew and A. di Silvio (1914).

This is a remarkable and spectacular statement, because it is the first time in history that I know, that a person reported an experiment, and then considered errors, not just the results.

## ARISTOTLE AND GALILEO

Philosophers today view Aristotle's contribution to knowledge somewhat as follows:

> *"Aristotle's creation of a new discipline of thought remains among the lasting achievements of the human mind. It may be doubted if any other thinker has contributed so much to the enlightenment of the world. Every later age has drawn upon Aristotle and stood upon his shoulders to see the truth."*
>
> —Will Durant

However in Aristotle's writings on physics he proposed and people believed a whole lot of nonsense. As Prof. Whewell in his *History of Inductive Sciences* says:

> *"There is no fundamental truth in the knowledge of science today which can be credited to Aristotle and his followers. Also there seems to be no physical doctrine today of which any anticipation can be credited to Aristotle. Yet Aristotle is still considered by many to have been the world's greatest collector and systematizer of knowledge."*
>
> — H.P. Girvin *A Historical Appraisal of Mechanics (1948)*

These are the two different opinions of Aristotle, the philosopher's and the scientist's. Chapter 10 continues our story of the role of Aristotle in the birth of science.

## THE NEXT REVOLUTION - THE EARTH'S HISTORY

If you would have run into that grouch, Sir Isaac Newton, in his old age wandering the West End of London and would have asked him what he was up to in his scientific studies he may have said

> I'm not doing anything on calculus or other boring subjects like that. Instead I'm deep into a very exciting and important project. I'm using all my mathematical powers to study the history of the earth. My friend Prelate Ussher of Ireland (Ireland's chief churchman) has determined the age of the earth by studying the Bible, going back in time from the birth of Jesus, from son to father and so on, and he determined that

> the earth was created at sunset (5: 45 PM) on 23 October, 4004
> BC. Our church has accepted this date. Well, I am checking the
> calculations and I find that his date is about 50 years
> off from the correct value. I will publish my
> results after I have checked them.

Isn't it amazing that until about 150 years ago churchmen, scholars and people thought that the earth was created about 6000 years ago? Then, in 1859 Charles Darwin came out with a book which indicated that the world was much, much older, and also told how life evolved on earth. This was a revolutionary idea which was attacked violently.

**Man does not accept new ideas easily, and even today we find objectors to Darwin's ideas.**

## REVOLUTION IN PHYSICS - THE THEORY OF SPECIAL RELATIVITY

When you run against the wind, or with the wind behind you, your speed will differ. Light was supposed to travel through a mysterious fluid called the ether, and its speed too should differ when travelling in different directions through the ether. Knowing how our world flows through the ether will tell how fast our earth is moving through the universe—a most important physical measure. The earth goes around the sun at 18 miles/sec and so we should be able to tell how fast we are flowing through the ether by measuring the 'ether wind'.

Two American scientists made very precise experiments, but to their surprise and dismay could not find any difference in the speed of light. For 26 years physicists puzzled over this until a young man, Albert, said:

> *"OK, I will accept what has been measured, that the speed of light is the same no matter in which direction it goes."*

Then he looked at the consequences of this and came up with the famous relativity equation $E = mc^2$

This reminds us of the empiricist's guiding rule: If your experimental result does not fit your theory then check your experiment. If it is OK then look at your theory and find its flaw - because theory must agree with your experiment.

Einstein's conclusion that some stuff actually disappears or is actually created was unbelievable. Again, this was difficult to get accepted and a quarter of a century after Einstein published his work on relativity a book was published in Germany called:

> *"One Hundred Authors Against Einstein"*
> See a review of this book by
> von Brunn in *Die Naturwissenschaften, 11, 254-6 (1931)*

## REVOLUTION IN EARTH SCIENCES - WEGENER'S MOVING CONTINENTS

In 1915 Alfred Wegener, a meteorologist, proposed that the earth's surfaces were not static but slowly moved as if they were floating on very thick treacle. He was attacked most viciously with scientists calling him all sorts of nasty names. It took until 1971 for the U.S. geological societies to finally accept this idea and today nearly all accept this view.

## THE DIFFICULTY OF GETTING A NEW IDEA ACCEPTED

**Galileo 1636 to 1992**  Against the accepted view of the church of Rome which is said to report to us God's word, Galileo proposed his new idea that the sun was the center of the universe, and not the earth. This was heresy. It was against church teachings and thus against God's word. His book was banned and all copies that could be retrieved were burned. Only in 1992 did the Pope finally say that Galileo was not wrong.  It took over 350 years to get this new idea truly accepted.

**Darwin 1859 until today**  Some scholars feel that God designed all animals in his workshop and any thought that they could evolve and change is sacreligious and against church teachings. Darwin disagreed. Today some Midwest states in the U.S. are protesting the teaching of Darwin's revolutionary ideas. Imagine the head of the department of philosophy of an American State University marching in front of the State Capitol with a placard demanding that state schools teach these creationist ideas in science classes?

## Summary

**Wegener 1915 -1971** This idea of moving continents does not counter any religious teachings, it just violates our common sense. It took 55 years for American geologists to change their accepted views in light of evidence.

**To summarize the ideas in this chapter let me quote Richard Feynman:**

> *"Experiment and observation are the sole and ultimate judge of the truth of an idea. It is not philosophy that we are after, but the behavior of real things . . . I like science because when you think of something you can check it by experiment; 'yes' or 'no'. Nature says, and you can go on from there progressively. Other wisdom has no equally certain way of separating truth from falsehood."*
>
> —from *The Letters of Richard P. Feynman, (2005)*
> by Michelle Feynman

# CHAPTER 2

# TRUTH AND KNOWLEDGE

*Philosophy has been defined as*
*"an unusually obstinate attempt to think clearly."*
*Bertrand Russell prefers to say that it is*
*"an unusually ingenious attempt to think fallaciously."*

Mankind has come up with all sorts of stories about our world and our past. These stories coagulate into myths, legends, and beliefs, for example: Kikuyu truth, Inuit truth, Navajo truth, Yanomamo truth, Kung San truth, Islamic truths ~ Shiite or Sunni, Hindu truth, the many, many Christian truths, and so on. Yes, man has come up with an endless number of self-consistent networks of myths which he believes are true statements.

Like an infectious disease these myths spread in populations, become more complex and some eventually die. Others are overwhelmed by "more powerful" myths.

It is interesting that just about every human is affected by and believes some myth or other, and it also is curious that when a myth consumes a person he often becomes intolerant of people believing some other myth, so intolerant that he may refuse to have lunch with the other person, or may even go to war or burn the one who adopts a different myth. This is the curious power of myths.

**For example consider the ideas of how old the earth is**

<u>The Young Earth</u>   Here are some proposals. <u>Prelate James Ussher</u> of the Protestant Church of Ireland in 1650 AD deduced from the Bible that the universe and our world were created at nightfall on Sunday, the 23rd of October, 4004 BC.

<u>Venerable Bede</u> said that creation occurred a few years later, on 3952 BC, while <u>Scaliger</u> said that it occurred on 3949 BC. <u>Isaac Newton</u> and still others proposed closely similar dates. All these studies relied on the ultimate authority, the Hebrew Bible (the Old Testament). Many people believe these stories today including the majority of Americans. These are **<u>creationist</u>** views.

**A Very, Very Ancient Earth**  At the other extreme is a Scandinavian legend which says

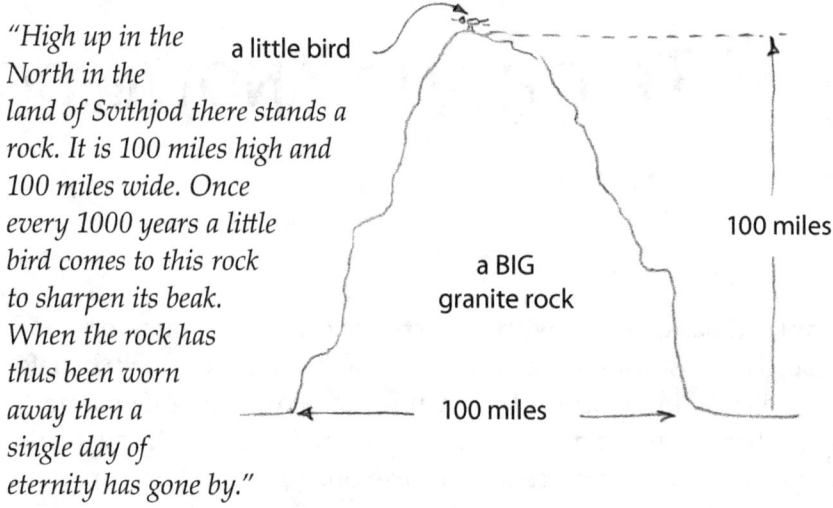

*"High up in the North in the land of Svithjod there stands a rock. It is 100 miles high and 100 miles wide. Once every 1000 years a little bird comes to this rock to sharpen its beak. When the rock has thus been worn away then a single day of eternity has gone by."*

—from Hendrik van Loon
*The Story of Mankind,* 1922

**From Science** with Darwin's theory of evolution, with its supporting physical evidence, we learn today that the universe was created over 15,000 million years ago, and the Earth a mere 4,600 million years ago.

Today, most Americans believe, at the very same time, two contradictory truths - the Creationist and the Scientific. If our earth's history was written at a rate of 100 years per page how thick would our history be according to these truths?

In the **Creationist** view the history of the earth would fit comfortably in a 60 page paperback, while the **Scientific** view would need a bookshelf close to two miles long to fit its needed volumes. This difference does not represent a minor disagreement. One cannot go around accepting the consequences of one of these truths in some places, and the other, in other places.

**Galileo was the pioneer who led the way to the building of the consistent network of statements which became science.**

Let us sketch some of the consistent networks of statements which are believed to be true by one or other groups of people

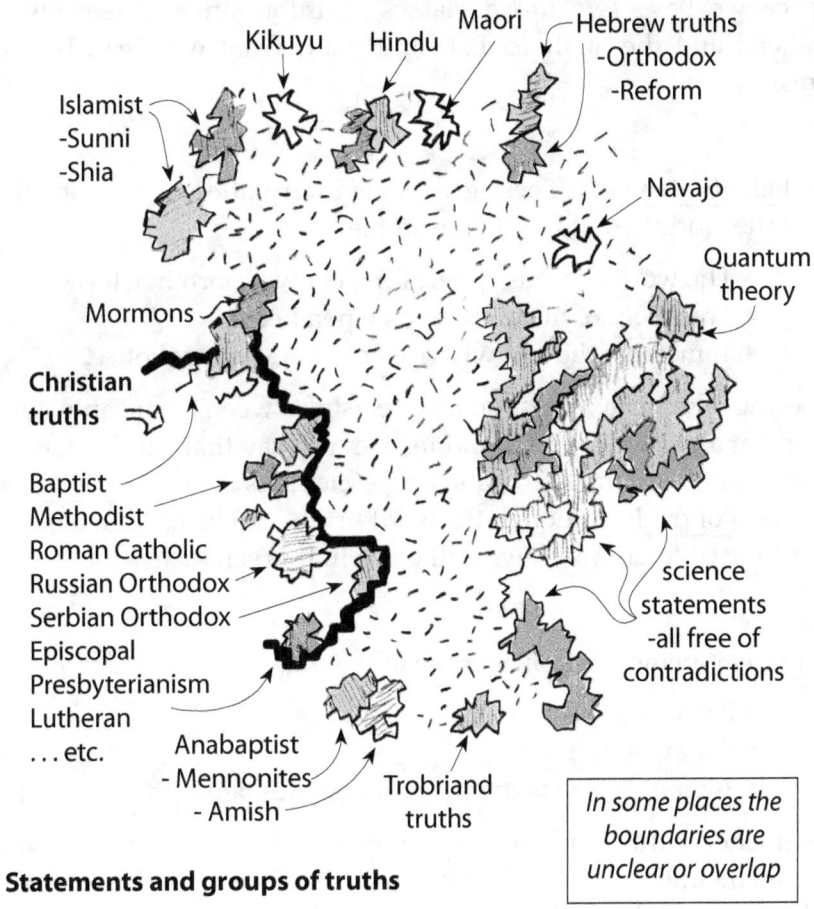

**Statements and groups of truths**

*In some places the boundaries are unclear or overlap*

## STATEMENTS AND KNOWLEDGE

Overall, man's three distinct activities are <u>science</u> (the attempt to understand the world about him which enables him to predict, hence control future events), <u>ethics and religion</u> in the broadest sense (his concern with what is right and wrong, how he ought and ought not to behave), and <u>esthetics and art</u> (his appreciation of beauty). Much confusion may arise from trying to deal with statements in science, art and religion all at one time, by making sweeping generalizations about all three and by judging ideas in these different areas to be alike. For example knowing how people act (empirical knowledge) doesn't tell us how they should act (ethics).

In this volume we will limit ourselves to discussions about science, and within it to some questions related to how we acquire knowledge. In science we have two broad classes of informative statements, the synthetic and the analytic. Let us explain what we mean by these terms.

**Synthetic Statements**. These give some information or say something about the world around us. For example:

- The world famous Albert Einstein was born in Chicago
- This piece of metal expands when heated.
- Tomorrow the sun will rise at 6:14 AM at my house.

By seeing whether a correspondence exists between the meaning of the statement and the world around us, we may say that such a statement is true, likely to be true, probably false, etc. This is the correspondence definition of truth. Although the word 'truth' can be used in a number of entirely different ways, we will use it in this sense.

**Analytic Statements.** Consider the following:

- Bachelors are unmarried men
- $X = 2Y$
- A straight line is the shortest distance between two points

Statements of this kind are not very useful, they have no truth value, they are definitions and in a very proper sense are mental creations. As such there is no limitation to what we may dream up or what mental games we may play with such statements. Pure mathematics is the best example of this activity. All this time the mathematician is up in the clouds, in a world of fantasy, playing with analytic statements. The question of truth just does not appear. Doesn't it remind you of Alice's adventures in Wonderland? Is it at all surprising that such flights of the imagination were created by a mathematician-logician?

Finally, we should point out that discussions and conversations use many other types of statements: imperative (turn left), exclamatory (wow, that was great) and emotive (what a beautiful sunset). These do not concern us.

## Problems

a) *Which of the following statements are informative?* These communicate thoughts or ideas or pass on acquired knowledge. *And which are not*, and are emotive, exclamatory, imperative, etc. ?

b) If informative, are they synthetic (true or false), analytic, or nonsensical?

<u>Note</u> that some of these statements can be classified in more than one way.

1. My friend, Don Pettit is in Antarctica right now.
2. Shut the door.
3. I wish it were Saturday.
4. Dr. Kinsey says that it is good for man to have two wives.
5. It is good for men to have two wives.
6. An electroencabulator produces electricity at 3.5 mil/kWh.
7. There is one girl in this class.
8. We have weekly quizzes.
9. We hope that weekly quizzes will be eliminated.
10. Stop giving weekly quizzes.
11. Weekly quizzes - ugh!
12. Do we have weekly quizzes?
13. Dogs like to roll on the grass.
14. 8 times 8 equals sixty-five.
15. Caesar is a prime number.
16. God is all powerful.
17. We hate the cruel, unjust, torturing weekly quizzes.

# Chapter 3

# Reason, Logic and Deduction

Now, people do not go about uttering or thinking isolated statements. Certain statements bring to mind other statements. This "bringing to mind" is called making an inference, or making an argument. This ability of inferring is called reason. Pierre-Simon Laplace, the great French philosopher-scientist, thought this so uniquely an ability of man that he called man "l'animal raisonable", the animal capable of reason.

As we may expect, one may reason correctly or incorrectly. But how to tell what is a correct inference and what is not? That question is the subject of the study called logic. So, logic is the study of argument, or how to reason correctly.

## Logical analysis

If we wish to evaluate for soundness a whole series of inferences, one depending on the other, we must first unscramble the discussion into simple arguments, each containing one inference, and for each argument we have to consider what are the statements used as evidence, and what are the conclusions reached from this evidence. This procedure is called logical analysis. Since many arguments are jumbled or omit statements, or take them for granted, this procedure is sometimes not that easy to do, and at times is well nigh impossible, as can be shown with many political speeches and philosophical essays.

After the individual arguments have been examined we may consider the argument as a whole. This is called a complex argument . As you may expect, this can become quite involved, so we will only touch what we consider to be its most important aspects.

## Simple arguments

When we know what has been given (the premises), and what is inferred (the conclusion), we notice that inferences fall into two categories, deductive or inductive, depending on whether the conclusion necessarily follows or only probably follows from the evidence.

**Deduction** is the name for a necessary argument, and we call a properly made deductive argument a valid argument. In a deduction the conclusion Q necessarily follows from the premises P1, P2, etc. which are the statements set forth as evidence,

$$P1, P2, \ldots \{\text{all lead to}\} \ldots Q$$

The words valid and fallacious concern the inference of a deductive argument. The argument is valid if the inference is correctly made, and invalid if the inference is incorrectly made. The words true and false do not concern the inference, but only concern the statements, and whether they agree with experience.

### Example of a valid deductive argument

- P1   Fish are flying creatures  (false)
- P2   Flying creatures wear goggles  (false)
- Q    Therefore fish wear goggles  (false, but valid)

**Induction** is the term to describe a probable argument, and the term cogent is used to describe what we consider to be a pretty good or a convincing inductive argument. The term valid is not used to describe inductive arguments.

These arguments involve synthetic statements, never analytic, *see Chapter 2*. In an inductive argument it is claimed that the evidence is sufficient to make the conclusions, at the very least, more likely to be true than false. The general form of the argument is "if P then most probably Q", or

$$P \ldots \left\{ \begin{array}{c} \text{tends to establish} \\ \text{seems reasonable to suppose} \\ \text{goes to show} \end{array} \right\} \ldots Q$$

In induction we do not use the words valid and fallacious to describe the argument. The conclusions are always probable, although some may be more likely than others.

**Example of an inductive argument**
- P1   Bear number 1 is brown (true).
- P2   Bear number 2 is brown (true).
- P3, P4, P5   and so are numbers 3, 4 and 5 (true).
- Q   Therefore I conclude that all bears are brown (?).

The distinction between <u>necessary</u> and <u>probable</u> is perhaps the most fundamental way of classifying arguments. The rest of this chapter deals primarily with deduction. Induction and statistics (the systematic application of induction) will be taken up in Chapters 5, 6 and 7.

## COMPLEX ARGUMENTS

We have seen above how a simple argument may be considered to be inductive or deductive depending on whether the conclusion probably or necessarily follows from the statements taken as data for the inference.

Let us now consider a complex argument where the conclusion of the first inference is taken as the starting point for the second inference, and so on. Is the argument as a whole (the complex argument) inductive or deductive? With a little reflection you should agree with me that in a complex argument:

a. If all the individual arguments are deductive then we can consider the conclusion of the argument as a whole to be deductive.

b. If one or more of the individual arguments in the chain of arguments is inductive then the conclusion of the whole chain of arguments only probably follows, hence the whole argument is inductive.

## THOUGHTS ON THE NATURE OF DEDUCTION

Consider a one step deduction, called the **syllogism**. If a person asserts or accepts as true the two statements P1 and P2, then based on this alone how is it that he can assert as true a particular third statement, and this without getting out of his armchair to examine the facts? What justification has he for doing this and what is to guarantee that he won't be wrong in doing this? I suppose that the stages in man's thinking which led to this remarkable situation were somewhat as follows:

> At first, if man of long, long ago would be asked the truth of Q given that P1 and P2 are true he would go out and check the facts about Q to decide its truth. The information about P1 and P2 would not be used. However he may have a vague feeling that somehow the P's may be related to Q. It must have dawned on him that some combination of P1 and P2, both true, always gives true Q.

Actually the earliest writings which show this awareness are the writings of Thales of Miletus in the 6th century B.C.

Two hundred years passed before the anatomy of deductive argument was laid open by Aristotle in the first work devoted to this subject. Here he brought together and summarized the rules of correct deductive reasoning known at that time. He showed that what was important was not the meaning of the statements themselves but the form of the argument. He generalized, essentially presenting these rules of correct deduction.

After 2000 years of virtual stagnation, an Englishman, George Boole, made a grand contribution by discovering a link between logic and mathematics. He showed that the laws given by Aristotle could all be considered to make up a mathematical system.

Finally, in deduction we may want to represent statements and their relationships, the syllogisms, by diagrams, called Venn diagrams. This simplifies the solving of problems. Here are two examples of Venn diagrams, one valid, the other not valid:

*Venn Diagrams:*

| Valid | Not valid |
|---|---|
| P1 ... Students are Human | P1 .. Students are Humans |
| P2... Humans are Animals | P2 .. Students are Animals |
| Q ... thus Students are Animals | Q... thus Humans are Animals |

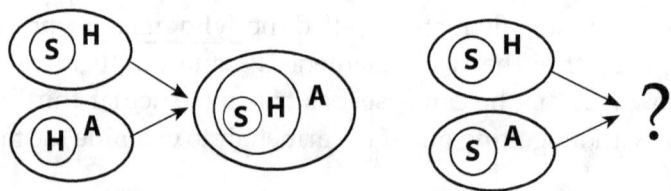

Let us look at the general form of the syllogism. For this, consider three groups of people: X, Y, and Z and let the symbols 'X are Y' mean that all the people in group X are part of group Y.

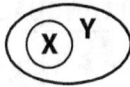

Practice Exercise. Here are four possible argument forms for X, Y and Z:

|  | (a) | (b) | (c) | (d) |
|---|---|---|---|---|
| for P1 ... | X are Y | X are Y | X are Z | X are Y |
| for P2 ... | Y are Z | X are Z | Y are Z | Y are Z |
| and Q follows: | X are Z | Y are Z | X are Y | Z are X |

Try to display these four arguments with Venn diagrams, and with them, tell which argument forms represent valid conclusions.

Argument from (a) is valid; (b), (c) and (d) are not.

## COMMENTS

1. A valid deductive argument form always gives a true conclusion if you start with true propositions.

2. If the conclusion of a valid deductive argument form is false then at least one of the propositions is false.

3. A valid deductive argument from false statements (or propositions) can give a true conclusion.

4. Beware, some arguments have hidden or unstated premises.

5. Generalizations (laws) and explanations (hypotheses or theories) are all inductions and have no guarantee of certainty.

6. Deductive inferences from analytic statements follow with certainty. However the conclusions say nothing about the world.

7. Modern science exhibits the remarkable union of these two types of arguments, deduction and induction. This is called the hypothetico-deductive method. Here we start with both types of statements, synthetic and analytic, but always end up with synthetic statements.

8. Invalid deductive arguments can give true or false conclusions.

## DIFFICULTIES WITH THE ENGLISH LANGUAGE

English is an awkward language which has some logical ambiguities and difficulties, particularly when it comes to negative statements. Interestingly, many other languages seem to avoid these problems. For example:

a) If you hear the following: "You're not going to school today, are you?" What does the answer 'No' mean?

b) If someone says: "A or B is OK". Does it mean that both are OK (inclusive), or does it mean that only one or other is OK (exclusive)?

If we are aware of these weaknesses we can often avoid them. Also, sometimes the meaning of a statement depends on the inflection of voice of the speaker.

Here are some useful tricks and relationships when dealing with negatives. In addition, in working with the graphical aid, the Venn diagrams, let the symbol $\overline{A}$ stand for everything that is not **A**. Then we have the following equivalences

Only **A** are **B** = $(B(A)$ = $(B)(\overline{A})$ = $(\overline{A})\overline{B}$

Only **A** are $\overline{B}$ = $(\overline{B}(A)$ = $(B)(\overline{A})$ = $B(\overline{A})$

Only $\overline{B}$ are **A** = $(A)\overline{B}$ = $(A)(B)$ = $\overline{A}(B)$

Can you represent the statements above in terms of Boys and Girls where

$A$ = boys, $\overline{A}$ = girls, $B$ = are clever, $\overline{B}$ = are not clever?

## QUESTIONS

*Are the following statements true or false?* Write out the correct statement to each one which is false. Note, do not change the underlined portion of the statements given:

1. **If the conclusion of a valid deduction is true,** then the premises must be true.

2. **In a valid deduction if the premises are true then** the conclusion may be true.

3. **Induction is the name for** necessary arguments.

4. **Deduction is the name for** both necessary and probable arguments.

5. **Logic is the study of argument** where argument is any discussion containing inferences.

6. **Any kind of statement can be used as evidence** in an inductive argument.

7. **A simple argument never contains more than one** deduction or one induction, but sometimes contains both.

8. **A complex argument is inductive** only if all the inferences in the argument are inductive.

9. **The word valid applies to** a "good" inductive argument; **the word cogent applies to** a bad inductive argument.

10. **The word valid applies to** a "good' deductive argument while **the word cogent applies to** a 'bad' deductive argument.

11. **In a valid deduction if the premises are false then** the conclusion must be false.

12. **In a valid deduction if the conclusion is not true then** not all the premises are true.

13. **Deduction is the name for** simple arguments.

14. **Induction is the name for** complex arguments.

15. **Logic studies whether** statements are true or false.

16. **Only synthetic statements can be used as evidence** in a deductive argument.

17. **Logic is used** only in the fields of knowledge, science and mathematics.

18. **A complex argument is deductive** if at least one inference is deductive.

19. **The word valid applies to** a "good' deductive argument **while the word cogent** applies to a 'bad' deductive argument.

20. **Algebra deals with** synthetic statements.

21. **Valid is the name for** true deductive arguments.

22. **A deductive inference is one which** must lead to true conclusions.

23. **A simple argument is one which** you can grasp right away.

24. **A complex argument is one which** is difficult to follow.

25. **The words true or false refer to** "good' or "bad" arguments.

26. **A complex argument is inductive if it** has no deductive inferences.

27. **In an invalid deduction from true** premises the conclusion is always false.

28. **If the conclusion of a valid deduction is true** then the premises must be true.

29. **In a valid deduction if the premises are true then** the conclusion must be true.

## Shorter Problems

*What can you say about the following: valid, invalid, cogent or nonsensical? And is any statement missing in the argument?*

41. Whales are mammals
    Whales suckle their young
    Therefore mammals suckle their young

42. Fish fly
    Those who wear goggles fly
    Thus fish wear goggles.

43. Is the boyfriend a gentleman? Yes, because gentlemen prefer blondes and he prefers me, and I am a blonde.

44. Manitobans live in the northern part of North America, and Canadians live in the northern part of North America, therefore Manitobans are Canadians.

45. No cat likes dogs, but Kitty likes dogs. That means that Kitty is not a cat.

46. All pigs are greedy
    No pigs can fly.

47. Professor Rochefort's point is that all who apply themselves will succeed, but since you persistently neglect your studies your chances of success are zilch.

48. We ought to have socialized medicine in the US for it has worked well in Britain.

49. If intoxication follows from consuming whiskey and soda, brandy and soda, rum and soda, gin and soda, vodka and soda then obviously soda causes intoxication.

50. If a theory is true it will be confirmed by all our experiments. All our experiments have confirmed our theory, so it must be true.

51. Kant held that all proofs for the existence of God are fallacious. He was therefore an atheist.

52. You act just like every radical I have met. Need I say more?

**WHAT CONCLUSION CAN BE DRAWN FROM THE FOLLOWING PREMISES** - if you need them, use Venn diagrams to help you

61. No non-residents are citizens
    All non-citizens are non-voters.

62. Mammals are warm-blooded animals
    Cows are warm-blooded animals.

63. All puddings are nice
    This dish is a pudding
    No nice things are wholesome.

64. No one takes the "New York Times" unless he is well-educated
    No hedgehog can read
    Those who cannot read are not well-educated

65. Every person who is sane can do logic
    No lunatics are fit to serve on a jury
    None of your sons can do logic

66. Babies are illogical
    Nobody is despised who can manage a crocodile
    Illogical persons are despised

67. Only profound scholars can be Professors at our University
    No insensitive souls are great lovers of music
    No one whose soul is not sensitive can be a Don Juan
    There are no profound scholars who are not great lovers of music

68. All my sons are slim
    No child of mine is healthy who takes no exercise
    All gluttons who are children of mine are fat
    No daughter of mine takes any exercise.

69. When I work a logic example without grumbling, you may be sure that it is one that I can understand
    These examples are not arranged in regular order, like the examples I am used to
    No easy example ever makes my head ache
    I can't understand examples that are not arranged in regular order, like those I am used to
    I never grumble at an example unless it gives me a head ache.

70. No kitten that loves fish is unteachable
    No kitten without a tail will play with a gorilla
    Kittens with whiskers always love fish
    No teachable kitten has green eyes
    No kittens have tails unless they have whiskers

## Sound Reasoning?

Here is a group of arguments. Are they valid, invalid or what? It may be helpful to use Venn diagrams to solve the problems.

71. In talking about creatures I say:
    Every eagle can fly
    Some pigs cannot fly
    Thus some pigs are not eagles

72. About creatures I say:
    All lions are fierce
    Some lions do not drink coffee
    Thus some creatures that drink coffee are not fierce.

73. In considering candies I say:
    All these bon-bons are chocolate creams
    All these bon-bons are delicious
    Chocolate creams are delicious

74. Here's a tricky one:
    Sugar is sweet
    Salt is not sweet
    Thus salt is not sugar

75. Do you dream in technicolor? Now it is well known that only non-artists never dream in technicolor so you must be an artist.

76. If our theory is true then it will be confirmed by all our experiments. Well, it so happens that all our experiments agree with our theory, so the theory has to be true.

# Reference

Many of the problems in this chapter come from:

Lewis Carroll, *"Symbolic Logic, part 1, Elementary"*, Fourth edition, Dover Publications (1959)

# Chapter 4

# MATHEMATICS

*"What's one and one and one and one and one and
one and one and one and one and one?"
"I don't know" said Alice. "I lost count."
"Hah!, She can't do addition," said the Red Queen.
—Lewis Carroll*

## A. COUNTING

The basis of math is the act of counting, or relating the number of items in a group with a set of signs such as 1, 2, 3, 4, and so on. In some primitive cultures the people only had symbols for just two or three items. From then on they just said 'lots'. In such a society I wonder how a shepherd could keep track of his flock; and if someone stole a sheep how could he tell?

> **Not only man but animals and birds could count.** For example in James R. Newman's *The World of Mathematics*, pg. 433, we read of the lord of the castle who wanted to get rid of a noisy crow who flew to the castle's watch tower early every morning to make a wild racket there and thereby disturb his lordship's slumber. Whenever anyone entered the tower to shoot the bird with bow and arrow it stayed away. To deceive the clever bird, a brilliant plan was devised. Two men were sent into the watch tower, one left, while the other remained. But the crow counted and kept her distance. The next day three went and again she perceived that only two retired. It was found necessary to send five or six men in to the watch house to put her out in her calculation. **From this the lord inferred that crows could count up to four.**

## B. Theory of Numbers

The properties of numbers have been studied since ancient times. Here are a few examples:

The ancient Greeks said that a number could be **perfect, deficient** or **abundant,** depending on whether its proper divisors added up to it, or didn't, for example

- The divisors of 6 add up to $1 + 2 + 3 = 6$ ... so 6 is a perfect number
- The divisor of 9 add up to $1 + 3 = 4$ ... so 9 is a deficient number
- The divisors of 12 add up to $1 + 2 + 3 + 4 + 6 = 16$ ... so 12 is an abundant number

All perfect numbers (6, 28, etc.) found so far are EVEN. Are there any ODD ones? We don't know yet. If you can find one or can show that there are none, you will end up in mathematical history!

Pairs of numbers are considered to be **amicable** or **friendly** if the sum of the divisors of each one add up to the other. For example 284 and 220, a pair ascribed to Pythagorus, is amicable because

- the sum of the divisors of 220 add up to $1 + 2 + \ldots + 110 = 284$,

and  • the sum of the divisors of 284 add up to $1 + 2 + \ldots + 142 = 220$

Such pairs of numbers had a mystical aura, and superstition later maintained that two talismans bearing one and other of these numbers would seal perfect friendship between the wearers, or perfect love between two lovers. These numbers play an important role in magic, sorcery, astrology and the casting of horoscopes.

Some numbers are called **prime** because they have no divisors other than 1. Here are the first few:

1, 2, 3, 5, 7, 11, 13, 17, 19, 23, 29, 31, 37, 41, 43, 47, and so on. These primes seem to occur in pairs. How do you explain this curious phenomenon?

Also I challenge you to find the largest prime number in the world.

There are so many other fascinating curiosities with numbers. As a good source see Howard Eves *"An Introduction into the History of Mathematics"*, Saunders Series, 6th Edition, 1990 for an adventure into numbers.

## C. MATHEMATICAL SYSTEMS

Mathematics is very much like a game; however it is not one you can put away just because you may not like it. For it is a game you cannot escape in everyday living. You use it to pay bills, to remember how many friends you have, to count change, and to tell if you will be late for an appointment. However, not many of us know the basic rules of these games and it is precisely for that reason that we consider that subject difficult. As we shall see, mathematics is not just one game, but the name given to many different games, each one being called a <u>branch of mathematics</u>. But before we show you how to play these games we must introduce you to a few new terms.

In logic we talk about **informative** statements, statements about our world, which may be true or false.
- Einstein was a physicist (true)
- Copper is a friendly, furry animal (false)

How about the statement
- X is 2Y

Is it true or false? We cannot tell until X and Y are defined. This type of statement is called an **analytic** statement.

**Mathematics deals with analytic statements**

## D. NEW WORDS

A branch of mathematics has its equipment, called:

- its **elements** or the pieces we play with,

- its **postulates** or **axioms** which are the rules of the game, and

- its **theorems**, its deductive consequences, which are the allowed moves of the game.

As an example, in checkers the elements are the board plus the 24 pieces, the postulates are the rules that tell how a piece is allowed to move, and the theorems are the actual moves in a game.

## E. Different Kinds of Mathematical Proofs

In mathematics the deductive step, or the development of theorems, can come in a number of forms:

1. **Direct deduction** as shown in the syllogisms of the previous chapters.

2. **Mathematical induction**, an indirect proof. *See the problems in section N of this chapter for some examples.*

3. The **"reductio ad absurdum"** proof. Here you assume the alternative of what you want to prove, then show that this assumption leads to a contradiction. This shows that what you had originally assumed is wrong and that what you actually wanted to prove is correct.

Some mathematicians will not accept anything but the direct proof, so they reject whole sections of the subject, and tear out and stomp on those sections of mathematical textbooks which are based on those other kinds of proofs. Mathematicians are still arguing about these matters. Let us leave them to it.

## F. Development of a Branch of Pure Mathematics

**Step I**  We first decide what will be the elements, or the pieces of our 'game'. We will play with those pieces but we will not attempt to define them. They are just the pieces, that's all. All other terms that we may use later in our discussion must be defined in terms of these primitive and undefined starting terms.

**Step II**  We then state the rules to tell what we are allowed to do with these elements, and what moves we are allowed to make. These rules are called postulates. We can chose them in any way we wish, but with one condition; that they must not contradict one another, because if they do then the whole game becomes nonsensical.

**Step III**  With these postulates (rules for making moves) we may deduce certain new statements (make moves), and from them further statements (more moves) which we call theorems. In a sense we are playing the game.

Needless to say, since we started this chain of deductive inferences with analytic statements, the theorems must also consist of analytic statements.

Though games usually end after not too many moves it has been shown that for any branch of mathematics if one theorem can be deduced, then the "game" never ends and one can obtain theorem after theorem after theorem. How much more interesting are these mathematical games, always leading to new theorems, always bringing us new horizons.

## G. Pure Mathematics

We have shown how a branch of pure mathematics is developed. Let us now define pure mathematics. Pure mathematics is the sum of all the many branches of pure mathematics that have been made up or thought of. Let us dwell a while longer on this subject. Bertrand Russell in his monumental work "Principia Mathematica" wrote,

> "Mathematics is the science in which one never knows what one is talking about or whether what one says is true"

What did he mean? If we examine the three steps in the development of a branch of pure mathematics we see that we have been dealing all the time with analytic statements. So as the postulates are neither true or false, then the theorems deduced from them, which are also analytic statements, are neither true nor false.

## H. Applied Mathematics

Consider the three steps in the development of a branch of pure mathematics. Suppose we define the elements of **Step I** in terms of things we know so that the postulates of **Step II** are true for us. Then we can say that the theorems that follow are true. The discourse we get is then called a branch of applied mathematics. In many cases we can develop a number of branches of applied mathematics (sometimes called interpretations) from each branch of pure mathematics. As may be expected, applied mathematics is the sum of all branches of applied mathematics which may be obtained from all branches of pure mathematics.

## I. Mathematics and its Uses

What is mathematics? It's just the sum of all the branches of pure and applied mathematics that we have dreamed up.

Today mathematics is often divided into four large groups, depending on the types of elements and postulates involved. It should be understood that this classification is not at all rigid but merely a convenience. The groups are:

> **Arithmetic** - ambition, distraction, uglification, derision (according to the Mock Turtle, in *Alice's Adventures in Wonderland*)
> **Algebra**
> **Geometry**
> **Analysis**

In attempting to solve problems that may be met in the world we live in, man has found mathematics an indispensable mental tool. Its deductive machinery are the artillery drawn upon to solve practical problems from the counting of wives of yesterday to the production of cell phones of today. As we shall presently see, we cannot call it a science though it is an indispensable tool of science.

## J. Mathematics and Logic

Historically mathematics and logic developed independently as separate studies. However, the growth of both fields in modern times have caused each of them to invade territory traditionally belonging to the other. Thus we may ask if mathematics is part of logic or vice-versa. Those considering mathematics to be a part of logic say that logic considers reasoning in all forms, both deduction and induction, while mathematics is concerned with deduction alone. However, those considering logic as a part of mathematics say that the rules of correct reasoning can be written down in postulate form and can then be considered as a part of mathematics; hence the whole subject of logic is just one of the many branches of mathematics.

As questions such as these concern the foundation of these two subjects and have not as yet been answered to the satisfaction of all, we will leave them to the experts to argue about. Let us just illustrate these ideas with some examples of branches of mathematics.

## K. THE POSTULATE SETS FOR THE GEOMETRIES
### (three different kinds of geometry)

P1. Any two points can be joined by a straight line.

P2. A line can be continued in either direction for ever and ever.

P3. A circle may be described with any point as center and having any radius.

P4. All right angles are equal to one another.

P5a. One and only one line can be drawn through a given point parallel to a given line.

P5b. No line can be drawn through a point parallel to a given line.

P5c. Two or more lines can be drawn through a given point parallel to a given line.

a) An interpretation of the branch of mathematics based on P1, P2, P3, P4, P5a was given by Euclid over 2000 years ago and is called **Euclidian Geometry**. This geometry applies to a flat plate on which the three interior angles of a triangle add to 180 degrees.

b) The postulate set P1, P2, P3, P4, P5b is called **Riemannian or Elliptical Geometry**. This geometry applies to the surfaces of a sphere on which the interior angles of a triangle add to more than 180 degrees.

c) The Postulate set P1, P2, P3, P4, P5c is called **Lobachevskian or Hyperbolic Geometry**. This geometry applies to surfaces such as that of a trumpet horn on which the interior angles of a triangle add up to less than 180 degrees.

The last two branches of mathematics are known as **Non Euclidian Geometries**. Here are sketches of examples of such geometries.

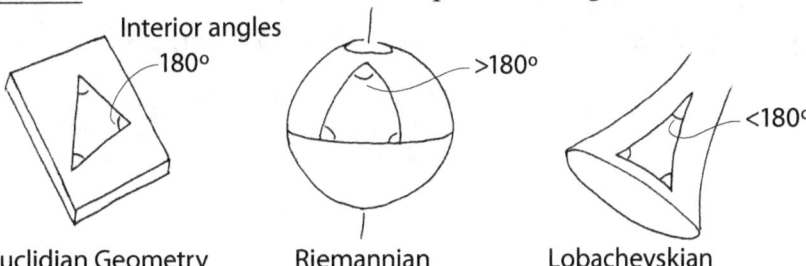

Euclidian Geometry     Riemannian     Lobachevskian

## L. Let us Develop our own New Branch of Mathematics

### a) Pure Mathematics

**Basis**

A set of elements a, b ... and a relation **R** connecting certain pairs of elements. If a is related to b then we shall write (a**R** b), otherwise we shall write (a**R̄** b).

**Postulates**

P1. If a and b are two distinct element, or in symbols, a ≠ b, then either (a**R** b) or (b**R** a), but not both

P2. If (a**R** b) then a ≠ b

P3. If (a**R** b) and (b**R** c) then we have (a**R** c) [**R** is transitive]

P4. We have exactly 4 elements

**Theorems and Definitions**

T1. If (a**R** b) then (b**R̄** a), [**R** is not symmetric]

T2. Given (a**R** b) and c ≠ a and c ≠ b, then we have either (a**R** c) or (c**R** b), or both.

T3. There is only one element 'a' such that (a**R**b) for all and every other element 'b' of the set [this is the uniqueness theorem]

D1. If we have (a**R** b) then by definition we shall say (b**D** a)

T4. If (a**D** b) and (b**D** c) then (a**D** c)

D2. If (a**R** b) and there is no element 'c' such that (a**R** c) and (c**R** b), then we may define and write (a**F** b)

T5. If (a**F** c) and (b**F** c) then a = b

T6. If (a**F** b) and (b**F** c) then (a**F̄** c)

D3. If (a**F** b) and (b**F** c) then we may define and write (a**G** c)

b) **Applied Mathematics** (3 branches)

**Interpretation 1 (geometrical)**
Elements - four distinct points on a line
 (a**R** b) - a is to the left of b

**Interpretation 2 (arithmetic)**
Elements - The four integers 1, 2, 3, and 4
 (a**R** b) - a is greater than b

**Interpretation 3 (a family)**
Elements - a man, his father, his father's father, and his father's father's father
(a**R** b) - a is the ancestor of b
(a**D** b) - a is the descendent of b
(a**F** b) - a is the father of b
(a**G** b) - a is the grandfather of b

## M. COMMENTS

a) Do these interpretations make sense to you? Use ordinary words to check them.

b) Devise at least one more interpretation.

c) Prove theorem T1, either
  - by a direct proof,
  - by an indirect proof (reductio ad absurdum).

d) Prove theorem T2. Note that in many of the early theorems of a branch of mathematics it is often easier to use an indirect proof. So let us try it here.

e) Some mathematicians do not like one or another of these postulates. For example, Korubski and his followers object to postulates of type P1, Brougher and his group object to P2 and Intuitionists object to P3.

f) In Aristotelians logic P1 is called the Law of Identity (that A is -A), P2 is called the Law of Excluded Middle (everything is either -A or $\overline{A}$), and P3 is called the Law of Contradiction (everything is A and/either $\overline{A}$).

## N. Problems which use Mathematical Induction

1. Prove that: $1^2 + 2^2 + \ldots + n^2 = \dfrac{n(n+1)(2n+1)}{6}$

Start the proof by showing that this equality applies to $n = 1$. Then assuming that it applies to any value $n$ show that it applies to $m = n + 1$

2. Consider figures made of a linear chain of triangles joined at their vertices, for example:

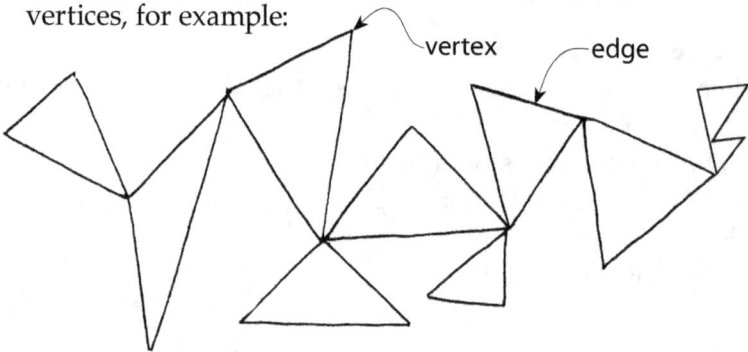

   a) Find the relationship between the number of edges E and the number of triangles T of such figures and prove that this relationship holds for all figures of this type.

   b) Repeat for the number of triangles T and the number of vertices V.

   c) Repeat for the number of vertices V and the number of edges E.

3. Consider the following limerick:

   *For a beach cabin or for a home*
   *Use tubing or Styrofoam*
   *Make it spherical - roughly*
   *With triangles - roughly*
   *And call it a geodesic dome.*

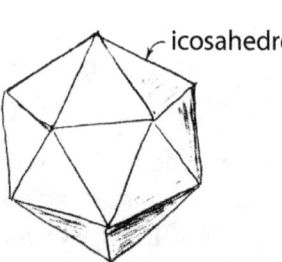

For a solid figure made up of triangular surfaces find the relationship between the number of edges E and the number of vertices V, and prove this relationship holds in general for any number of surfaces n.

## O. SOME INTERESTING PROBLEMS AND PUZZLES

**1. Here are some algebraic fallacies.**

### Addition

$$\begin{aligned} & 1 \text{ cat has 4 legs} \\ & \text{no ( i.e. 0 ) cat has 3 legs} \end{aligned}$$

Adding we get **1 cat has 7 legs**

### Multiplication

$$\begin{aligned} 2 \text{ pounds} &= 32 \text{ ounces} \\ 1/2 \text{ pound} &= 8 \text{ ounces} \end{aligned}$$

Multiplying we get **1 pound = 256 ounces**

### Division

| | |
|---|---|
| Assume that | $a = b$ |
| Multiply both sides by a, we get | $a^2 = ab$ |
| Subtracting $b^2$ from both sides, gives | $a^2 - b^2 = ab - b^2$ |
| Factoring both sides, we get | $(a + b)(a - b) = b(a - b)$ |
| Dividing both sides by $(a - b)$ gives | $a + b = b$ |
| If we take $a = b = 1$ we conclude that . . . | **2 = 1** |

### Inequalities

We must make use here of the following property of logarithms:
$$n.\log(m) = \log m^n$$

| | |
|---|---|
| We start with the inequality | $3 > 2$ |
| Multiply both sides by | $\log(\frac{1}{2})$ |
| Then | $3.\log(\frac{1}{2}) > 2.\log(\frac{1}{2})$ |
| Or | $\log(\frac{1}{2})^3 > \log(\frac{1}{2})^2$ |
| Whence | $(\frac{1}{2})^3 > (\frac{1}{2})^2$ or $\underline{\frac{1}{8} > \frac{1}{4}}$ ! |

2. <u>**Exercise in mental arithmetic**</u> - *no pencils or calculators, please.*

> Take 1000 and add 40 to it.
> Then add another 1000 for good luck.
> Now add 30 and still another 1000.
> To this add 20 and then another 1000.
> Finally finish this off by adding still 10 more.
> Did you get 5000?
> *Good for you, . . . but it's wrong.*

3. <u>**Only for good chess players**</u>

After 3 moves (W,B-W,B-W,B) here is the position of the board. How did we get to this crazy position? Can you get to this position after 4 moves?

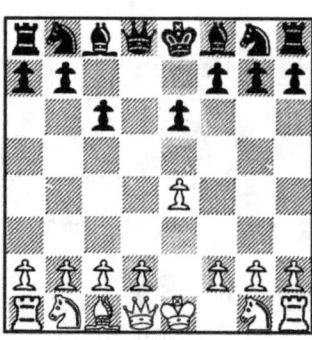

*Stefan Engstrom*

4. <u>**For ordinary chess players**</u>

Here are two end-games in both of which white moves and mates in two moves. Can you do it?

 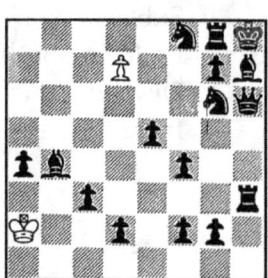

5. <u>**More chess**</u>

Place a rook in the top left hand corner of a chess board, and then move it once **and only once** into every square of the board and end up in the bottom right hand corner of the chess board. This is a tricky problem.

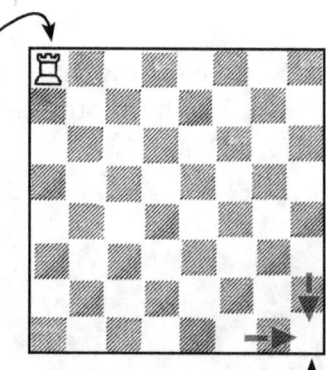

How do you do it?

## 6. The Map Problem

Given a map of the 48 states of the United States, could you color the map with three different colored crayons so that adjacent states are colored differently? Note that two states which meet at a point are considered to be not adjacent. Actually many years ago a mathematician proved that 5 colors could color any map that you can dream up so that adjacent states are colored differently. But to date no one has been able to draw a map that needs more than 4 colors. If you can come up with one you will become world famous. Try it.

- This is called the famous <u>four-color problem</u> and it has been teasing mathematicians for over 300 years.

## 7. The Game of Chinese Checkers.

A good player can transfer the 10 marbles from his home yard to the opposite yard in about 30 or 31 moves.

In *Scientific American* (Oct. 1976) Martin Gardner noted that *Ibidem* (Aug. 1969), a periodical on magic, published a proof that 29 moves represented the very minimum needed to do this, but then

a) What is the real minimum number of moves needed and can you find it?

b) If you can't find the minimum can you at least find the amazing 27 move strategy?

START with **ten** marbles

Elizabeth Wainscott pointed out that she could better it with 28 moves and then with an amazing 27 moves!!

**8. The Taiwanese Version of Chinese Checkers.**

On a smaller board having **6-marble** yards (see sketch below) Martin Gardner stated that one needed a least 18 moves to transfer one's marbles to the opposite yard.

If you've done that then try to do it in 17 moves. I can.

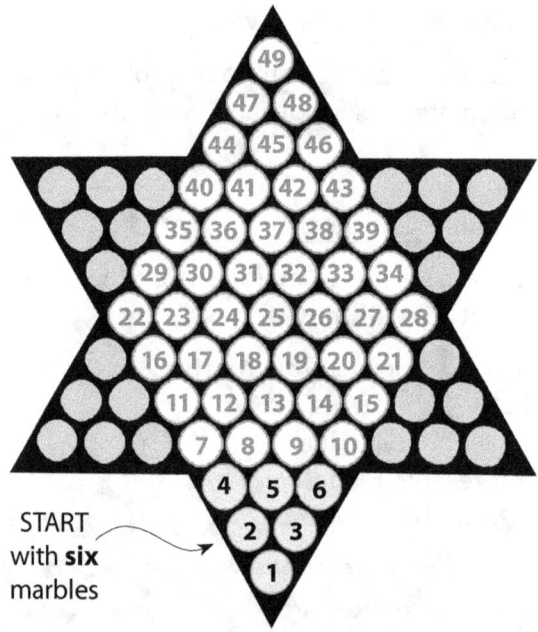

START with **six** marbles

**9. How to Match the World's Expert Chess Players.**

A, B and C are playing two simultaneous games of chess - A against B, and B against C. A and C are experts, B is a beginner. A has white in one game, B has white in the other game.

Show that with a clever strategy B can either win one game or else draw both games.

**10. Intuition versus Proof.** Answer the following questions intuitively, and then check your answer by calculation.

a) Each bacterium in a certain culture divides into two bacteria once a minute. If there are 20 million bacteria present at the end of t minutes when were there exactly 10 million bacteria present?

b) A grandfather clock strikes six in 5 seconds. How long will it take to strike twelve?

c) A bottle and cork together costs $1.10. If the bottle costs a dollar more than the cork how much does the cork cost?

d) Keith can do the job in 4 days, Skip can do it in 6 days. How long will it take the two of them to do the job together without interfering with each other?

e) My home is on a slight hill, so I ride my bicycle to my office at 6 mph, but I return home at 4 mph. What is my average speed?

f) Two jobs have the same starting salary of $12 000 per year. Job A offers an annual raise of $1200 while job B offers a semiannual raise of $400. Which is a better paying job?

g) Is a salary of 1 cent for the first half month, 2 cents for the second half month, 4 cents for the third half month and so, until the year is up, a good salary for the year?

11. **Hungry Cannibals.** A math teacher was captured by a tribe of hungry cannibals who say that he can make a statement and if it happens to be true then he will be boiled for dinner, but if false, then he will be roasted. What should the cannibals do if he says "Please, I insist that you roast me"?

12. **Integrate**  d(cabin) / cabin.

13. **A Dream**. Last night I had an unusual dream. In it I was a finalist in a TV quiz show. The MC announces my problem "Here are three boxes, A, B, and C. One box contains the tickets for a brand new Mercedes plus a one week vacation to the South Sea islands with the partner of your choice. The other two boxes are empty.

Which box do you choose?"

The boxes look alike, but since my girlfriend is called Alice I naturally choose Box A.

The MC walks to the boxes. However, before he opens Box A he gives me some extra information. (He is a tricky fellow). He opens Box B and shows me that it is empty. Now with this extra information it is time for me to choose.

a) Should I stay with my original choice of Box A or should I switch to the other unopened box, Box C

b) By the way, what is the probability of winning the prize if I stayed with Box A, or if I switched to Box C?

At this point I awake and felt that I just had to get the answers to these problems. Could you help me?

14. Find the diameter of the circle - without using calculus.

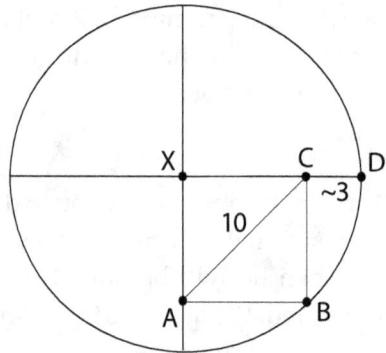

## REFERENCES

1. H. Eves *An Introduction to the History of Mathematics* 6th Ed., Saunders 1990.

2. F. J. Swetz *From Five Fingers to Infinity*, Open Court, Chicago 1994.

CHAPTER **5**

# INDUCTIVE REASONING

In general, in inductive arguments we have

a) Some statements (the evidence) which are presumed to be true, and
b) A statement which we think follows and which we think to be true, or likely to be true.

Unfortunately statistics has not been extended to help us draw conclusions for all problems. However it has been of great use in many scientific fields, so we will expand on it in Chapter 6. But first let us distinguish between two kinds of inductive reasoning.

## A. GENERALIZATIONS AND SCIENTIFIC LAWS

A generalization is an inductive conclusion which states that something is true about many or all members of a certain group. The evidence is always a repetition of that particular characteristic.

The evidence is not a variety of different fact, but the repetition of the same fact in different contexts. This is shown diagrammatically:

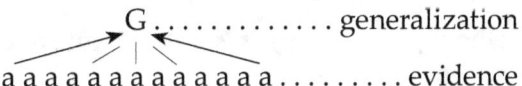
G ............. generalization
a a a a a a a a a a a a ......... evidence

In scientific fields generalizations are called scientific laws, and are expressions of some kind of uniformity.

Here are some generalizations and scientific laws:
- aspirin cures headaches.
- the ideal gas law relates the p, V and T of a batch of gas.
- men are taller than women.

- beagles make lousy pets,
- and so on

But in making generalizations we have some rather tricky matters to consider. For example consider the following two generalizations:

**Generalization 1**

At Yellowstone National Park the first bear I saw was a brown bear, the second bear that I saw was also brown, and so was the third, fourth and fifth. With this evidence I conclude that all Park bears are brown.

However in inductive inferences we must understand the meaning of the statements, not just the form of the argument. We must use good sense. Look at this:

**Generalization 2**

Intoxication follows from consuming whiskey and soda, rum and soda, and vodka and soda. Obviously you should cut out the soda.

And even worse still here is a story concerning the strictness of evidence:

> An engineer, physicist and mathematician were holidaying in Scotland. Glancing from the train window they observed a black sheep in the middle of a field. "How interesting" observed the engineer, " All Scottish sheep are black". The physicist responded, "No, no, you should have said that some Scottish sheep are black". The mathematician gazed heavenward in supplication and then intoned "In Scotland there exists at least one field, containing at least one sheep, at least one side of which is black".

**So be careful when generalizing. It is not a simple automatic process.**

## B. Hypotheses and Scientific Theories

An hypothesis is an inductive conclusion which proposes to explain why certain facts are as they are. The evidence is a variety of facts that converge on or suggest the hypothesis. Shown diagrammatically

# Hypothesis and Scientific Theories

```
        H . . . . . . . . . hypothesis
       ↗ ↑ ↖
 a b c d e f g h i . . . . evidence
```

Remember, all hypotheses are explanations

FACTS ⟶ give evidence for ⟶ HYPOTHESIS
⟵ gives explanation for ⟵

In scientific fields hypotheses are either called <u>scientific hypotheses</u> or <u>theories</u>, the latter being considered more reliable, probably true and more 'respectable' than the former.

Here are some examples of hypotheses and theories:

- Aristotle's theory that all feeling and all thinking comes from your heart, and that your brain has nothing to do with it.
- Copernicus' hypothesis that the sun is the center of the whole universe.
- Darwin's theory of evolution implies that the earth is very, very old.
- Einstein's theory of special relativity.
- Wegener's hypothesis that the continents wandered about the globe.
- Gall's hypothesis that the bump on your head can tell about your mental faculties, whether you are a genius, a singer, a lover and so on.

## SUMMARY

a) A generalization goes beyond the evidence to other things of the same kind, while a hypothesis makes a leap to something of a different kind.

b) In inductive reasoning we are never absolutely sure of our conclusions. The word 'probable' must always be used.

c) Statistics, when applicable, affords us the best and most reliable method with which to make inductive inferences.

d) The reliability of an inductive argument (its strength) is called its cogency.

e) In deductions if the conclusion follows from the premises we call the argument valid. The conclusion can be true or false, depending on the premises (see Chapters 3).

f) In induction we do not call the argument valid or invalid, we call it cogent if we think that its conclusion is probably true.

CHAPTER **6**

# PROBABILITY (P)

*We often find it difficult to make wise choices in situations which have uncertainty, and in no other branch of mathematics is it so easy, even for experts, to blunder as in probability theory.*

## LAW OF PROBABILITY FOR A SINGLE EVENT

Consider a situation where we have a total number "n" of possible outcomes called the sample space: some desirable, others not desirable. To get a better understanding how to find the probability of a situation you need to have an understanding of the idea of **sample space** and the idea of **outcomes.**

The **sample space,** S, refers to the total, or all possible outcomes of a random process that we are interested in; thus we have

S = 2 total outcomes for the flip of a coin; heads and tails

S = 6 outcomes for the throw of a die; 1, 2, 3, 4, 5, 6

S = 4 outcomes for the flip of two coins: when TH and HT are considered as separate outcomes.

The first figure on page 6.3 shows our sample space and all the outcomes in it (the dots). All these outcomes have the same probability whose total is unity.

The probability of getting a **desirable outcome** is

$$P = \frac{\text{the number of desirable outcomes}}{\text{total possible outcomes}}$$

### HERE ARE SOME EXAMPLES

1. If you flip a coin you have two possible outcomes, H or T. You choose either H or T, so the probability that your chosen outcome happens is

    $P$ = [H or T] / H + T = 1/2

2. If you throw a die you can get 1, 2, 3, 4, 5, or 6, each number being equally likely to appear, so the probability that your number comes up is $P = 1/6$ and the probability that an odd number of dots comes up is $P = 1/2$

3. If you throw two dice, one red and one black, you have 36 outcomes. The probability of getting two 5s in one throw is $P = 1/36$

4. The probability of getting at least one five from the two dice is $P = 11/36$

5. The probability of getting one and only one five with two dice is $P = 10/36 = 5/18$

6. You have two coins, a nickel and a dime, which I call "n" and "d" You toss them once. The outcomes possible are

   $H_n H_n \quad H_n T_d \quad H_d T_n \quad T_d T_d$

   These outcomes are all assumed to be equally likely, if $H_n T_d$ and $H_d T_n$ are considered to be separate events. So the probability of a given event occurring is $P = 1/4$

   If HT and TH are combined into one case this will lead to S = 3 outcomes with different probabilities, not 4. This is ugly. We do not deal with such cases in this book.

## RANDOM EXPERIMENTS A, B, ... AND THE SAMPLE SPACE, S

Consider experiments for which the outcomes cannot be predicted with certainty. These are called random experiments. The collection of all the outcomes of random experiments is called the outcome space, S.

If A is an event using outcomes, (objects or points in S), then A is a subset of S (number on a die). The same holds for B, which may have more or fewer points than has A (even number on a die). The symbols used in probability theory and the helpful Venn diagrams follow:

- S is the **sample space** with all its outcomes.
- All **outcomes** (dots) in a sample space have the same probability. These all sum up to $P = 1.0$.
- A is a set of elements which **is a subset** of S.
- A' is the **complement** of A (All elements in S that are not in A).

# Venn Diagrams

- A ∩ B is the **intersection** of A and B.
- A ∪ B is the **union** of A and B.
- A ⊃ B ⊃ S means that A is a **subset** of B, which in turn, is a **subset** of S.

## Graphs of Venn and Related Diagrams

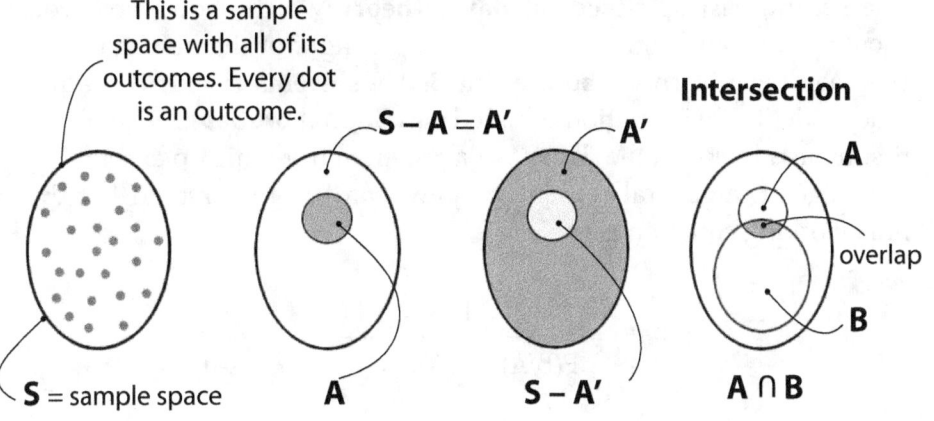

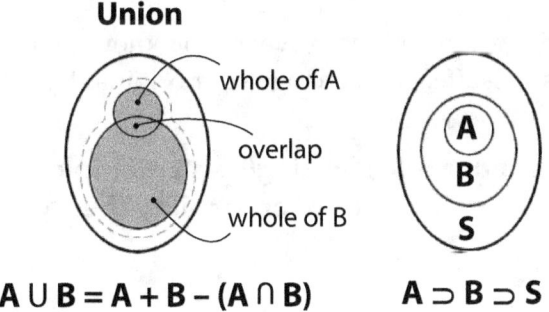

## 6.4

## LAWS OF PROBABILITY FOR TWO OR MORE EVENTS

There are two types of calculations to consider when dealing with the probability of two or more events.

**In the first case** the probability of A which you are interested in is not affected by any of your knowledge and probability that B has occurred. The two are independent of each other. So just add the two probabilities to get the joint probability of the two events This is called **independent probability.**

When A and B have no common points $P_{A+B} = P_A + P_B$

When A and B have some points in common $P_{A+B} = P_A + P_B - P_{AB}$

**The second case is based on Bayes theory** which shows how you should alter your original estimate of $P_A$ in light of the new evidence you may have learned about event B. Thus events A and B are not independent. This situation is called **conditional probability,** and was first treated by Reverend Thomas Bayes who wrote this up exactly 350 years ago. The general reasoning is awkward to understand but the final working equation is:

$$P(A/B) = \frac{P(B/A) \cdot P(A)}{P(B/A) \cdot P(A) + P(-B/-A) \cdot P(-A)}$$

P(A) when you know P(B)     P(B) when you know P(A)

So for multiple events A and B, where you know something about event B, P(B), this will affect and change your original P(A), (see the equation above).

**The above rules for single and multiple events are the basis of probability theory.**

Let us illustrate the use of conditional probability by solving a problem, see Example 1 as follows:

## Example 1 Using Bayes Theorem (see formula)

Bayes theorem tells how you should alter your original estimate of the probability of a person being a woman P(W) in light of the evidence you may have learned about long haired persons.

I am traveling by train from Moscow through Siberia to Vladivostok. At night, while resting in the upper bunk in my first class compartment, I am aware that someone is in the lower bunk. I wonder - is the person a man or a woman? I learn that that person is combing their long hair while preparing for sleep. Let me try to figure out if it is a woman or a man from the following information:

First I know that 75% of all travellers on this train are men, 25% are women. Also 80% of Russian women have long hair, but only 40% of Russian men do. *With the above information, what are my chances of finding that it is a woman in the lower bunk?*

$$W = \text{woman}$$
$$M = \text{man}$$
$$L = \text{long haired person}$$
$$P(W) = \text{probability that the person is a woman} = 0.25$$
$$P(M) = \text{probability that the person is a man} = 0.75$$
$$P(L/W) = \text{probability that Russian women have long hair} = 80\%$$
$$P(L/M) = \text{probability that Russian men have long hair} = 40\%$$
$$P(W/L) = \text{the person has long hair, therefore probably is a W}$$

From Bayes theorem, remember that you should alter your original estimate of P(L/W) in light of the new evidence you may have learned about long haired men, or P(L/M),

original knowledge of train travellers: $P(W) = 0.25$, $P(M) = 0.75$ ... new evidence of long haired Russians: $P(L/W) = 0.80$, $P(L/M) = 0.40$

So from Bayes theorem

$$P(W/L) = \frac{P(W) \cdot P(L/W)}{P(W) \cdot P(L/W) + P(M) \cdot P(L/M)}$$

Replacing terms gives

$$P(W/L) = \frac{0.25 \times 0.8}{0.25 \times 0.8 + 0.75 \times 0.4} = \frac{0.2}{0.2 + 0.3} = \underline{\mathbf{0.40}}$$

So the P is 40% that it is a woman in the lower bunk.

## Problems

Here are some problems in probability theory

1. How many outcomes of Heads and Tails can we get if we toss 3 coins?

2. How many outcomes can we get if we toss 4 coins.?

3. What is the probability of throwing a "2" and a "3" with one throw of two dice?

4. Which is more likely to occur in throwing three dice, a "6" or a "7"?

5. Four cards are to be dealt successively and at random without replacement from a deck of ordinary playing cards. What is the probability of receiving in order:

    a spade, a heart, a diamond and a club?

6. My blue box contains six marbles, 2 black and 4 white.

   a) I put my hand into the box and at random take out a marble - What is the probability that it will be black?

   b) I discard the marble. Then I put my hand in and pull out a second marble. What is the probability that it too is black?

   c) I put my hand into the box with its original 6 marbles and at random draw out two marbles. Both are black. What is the probability of that happening?

7. I have a green box that contains 5 black marbles and 3 white marbles. From the box draw out one marble. It is black.

   a) Discard the black marble and calculate $P_B$.

   b) Then pull a second marble from the box, and it is White. Find $P_W$.

   c) Return the 2 marbles to the original box. Then put your hand in and pull out at random two marbles, they are both black. Find $P_{2B}$.

8. I have two boxes one containing 2B, the other containing 1B and 1W marble. I choose a box at random and draw out a marble.
   a) What is the $P$ that it is B, or $P_B$?

   b) I return the marble to its box, and then take out two marbles using two separate random draws from one randomly chosen box. What is $P_{B+B}$?

9. Roulette has 38 pockets,"1 to 36" plus "0" and "00". I am told to bet on number 8, only 8, at $100 /play to win $3500. On the average how much will I win per play?

10. The birthday problem: How many people must you have in a group so there is a 50% chance of 2 of the group have the same birthday?

11. Another birthday problem: how many people must you have in a group so that some one of them has a 50% chance of having the same birthday that I have, which is July 6?

12. Which is greater, the number of 6 letter English words with an "n" as the fifth letter or the number of English words ending in "ing"?

13. Last night I put in 6 chocolates and 6 vanilla candy bars into my Chinese dragon jar. This evening I plan to pull out one bar at random.
    a) What would be the $\mathcal{P}$ of getting a chocolate bar?
    b) However I just learned that my friend Tom earlier this afternoon sneaked in and took two chocolates bars from the jar. How does that affect my probability?

14. This problem is a variation of an American TV show "Let's Make a Deal". In my version my friend Murty has two children:
    a) What is the probability that at least one of his two children is a boy?
    b) What's the probability that the other child is also a boy?
    c) If I hadn't met Murty and didn't know about his children what would my probability be if I guessed that he has two sons?
    d) I know that Murty's older child is named Ramkrishna. What's the probability that both of his children are boys?

15. When I toss two coins I can get 0, 1, or 2 heads. About 250 years ago the famous French mathematician, Jean d'Alembert, said that the probability of getting 0, 1, or 2 were each 1/3, but I think that he was wrong. What should he have said?

16. My box contains 4 marbles, 3B and 1W. I draw one at random.

a) What is the probability that it is B?

b) I then discard it. I then draw a second marble. What is the $P$ that it too is B?

c) I now put the 4 marbles in one box, and at random draw out two of them. What is the $P$ that both are black?

17. Last week Octave went to the State Fair in Salem where he saw all sorts of displays. He was intrigued by Harry the Huckster's booth which had a dirty green cloth covering his table. On the cloth were 26 big seashells placed upside down and labeled from A to Z.

    a) Harry beckoned Octave over to play his game. He said: "I've put a folded $20 bill under one of those shells. Come take a guess, it will only cost you $1." Octave was eager and said: "I'll choose shell A". What is his probability of winning?

    b) Harry said "I see that you've chosen shell A. You look like a good fellow so I'll give you a break. I'll turn over 23 of the other shells and show you that the $20 bill isn't hidden under any of them. Now let me ask you, would you like to keep your initial guess, shell A, or exchange your guess from shell A to one of the two still hidden shells? If there was no hanky-panky by Harry, then Octave using his math skills decides to stick with his initial guess, shell A. What would his probability of winning now be?

    c) And what would his probability of winning be if he switched from shell A to one of the two still hidden shells?

18. Harry also has 100 shells on another table, and under one of the shells he has hidden a precious 24 carat, old gold coin. Which shell should I turn over to have the best chance of retrieving the coin?

a) Of course I chose lucky #8. What would be my $P$ of winning?

b) But before I could touch #8 Harry said "Wait! 100 is too many shells so let me remove 90 of the useless ones and then ask if you want to keep your original bet or change your bet to one of the still hidden shells?'

19. Last night I put 6 chocolate and 6 vanilla candy bars in my previously empty purple jar.

    a) What is my probability of getting 2 chocolate and one vanilla bar in one grab?

    b) However I just learned that my good friend Tom, earlier this afternoon came into my office, saw the jar on my desk, and sneaked away two of those chocolate bars. So what is my new probability of getting those 3 bars?

    c) Then my daughter came later this afternoon and also took a vanilla bar. With only 9 bars left in the jar does that change my probability of getting those 3 bars (2 chocolate and 1 vanilla) in one grab?

20. On a table I have two face down card decks, one BLACK - A 2 3 4 5 6, the other RED - A 2 3 4 5 6. At random I pull out one card from each deck, turn it over and lay it on its deck. What is the $P$ that:
    a) two '4's show
    b) no '4' shows
    c) only one '4' shows

21. Combine the cards of problem 20 into a shuffled 12 card deck. Then pull out two cards, one at a time, and turn them over. What is the $P$ that:

a) two '4's show
b) no '4' shows
c) AT LEAST one '4' shows
d) only one '4' shows

22. Two coins, a penny and a nickel, are dropped into aa box divided in 2 equal sections and they do not touch each other. The coins are marked **H**ead on one side and **T**ail on the other.

    a) How many arrangements of one coin are there in the box?

    b) How many arrangements of two coins are there in the box?

23. Repeat problem 22 but with 3 coins, a penny, P, nickel, N and a dime, D.

24. Keith goes to work either by bus or by car. If he goes by car there is a 50% chance that he will arrive late for his meeting. If he goes by bus he will be late only 10% of the time. Keith was late last Monday. Did he drive his car or ride the bus?

    a) Because we have no idea which method he prefers we guess that 50% of the time he takes the bus and 50% of the time he drives. With this new information, what is the probability that he drove to work that day?

    b) Bekki, his wife, tells me that about 10% of the time he drives to work. What is the the new probability that he drove to work that day?

25. Throw two dice. What is $P$ that no "6" shows up?

26. Throw two dice. What is $P$ that both "6"s show up?

27. Throw two dice. What is $P$ that only one "6" shows up?

28. NIH has announced that a very unusual disease, call it X, has invaded the US: where from, we do not know, but once you get it you become cross-eyed, your ears curl and you die. Luckily there is one simple cure - a chicken-liver-mushroom preparation, call it Y, which your doctor can prescribe cheaply from the local Russian delicatessen.

    If you have X, the prescription Y will cure your disease completely; however, if you don't have X, then Y is harmless to 99% of people but 1% are affected adversely and may die.

    The doctor says that so far only 300 people in the US have been infected by this curious disease X.

    Should I take this medicine or not?

29. With the data given in Example 1 of this chapter, what are my chances of finding a man in the lower bunk of my compartment according to the equation from Bayes theorem?

Delightful References:

*Chance Rules*, Brian Everitt, Second Edition, Springer, 2008.

*The Drunkard's Walk: How Randomness Rules Our Lives*, Leonard Mlodinow, First Vintage Books Edition, May 2009.

CHAPTER 7

# STATISTICS

*A man with one watch knows what time it is,
the man with two watches is never sure.*

The word "statistics" has two somewhat different meanings

1. The word <u>statistic</u> refers to a single measurement of what we are interested in (his height is 5'4"), and in this sense the word statistics refers to the collection, tabulation and description of a collection of such data.

2. In the singular sense, the word <u>statistics</u> is a method for drawing inferences about a population of data from a sample of data taken from this population. When applicable, statistics gives us the best and most reliable way of making inductive inferences, or for saying something about the population from the data of a sample.

In effect then there are two meanings to the word "statistics" and we can ask two questions about a sample.

1. How can we describe the sample usefully and clearly?

2. From the evidence of the sample how can we best infer conclusions about the population, and how reliable are these conclusions?

**Question 1 • How to describe the sample**

The subject matter of descriptive statistics uses words such as: *tables, charts, histograms, geometric mean, harmonic means, modes, medians, deciles, standard deviations, skewness*. This information must be used with care if it is not to be deceptive.

**Example** Here are the monthly salaries paid by a small company to its eleven employees

$ 2000
$ 2500
$ 3000
$ 3000 ←——— mode (most popular value)
$ 3000
$ 3250 ←——— median (midpoint)
$ 4000
$ 5000
$ 6500
$ 7500 ←——— mean (arithmetic mean)
$ 50 000

To show that the boss is stingy, what measure would the employees promote? However to show that he is generous, what measure would the boss point to?

**Question 2 • What can you say about the population from the data of the sample.**

*This is the subject of modern statistics, and this is what we deal with in this chapter.*

**The terms we use in statistics:**

An <u>observation</u> is a recording of data.

A <u>population</u> is the whole set of observations or potential observations of size W.

A <u>sample</u> is a part of the population.

The <u>size n of a sample</u> is the number of observations in the sample.

A <u>random sample</u> is one drawn from a population is such a way that all potential observations have the same chance of being chosen. In this chapter when we say "sample" we will always mean "random sample".

A <u>parameter</u> is a property of the population. (The average height of all college basketball players in the U.S. is 6' 2").

A <u>statistic</u> is a property of the sample. (The average height of our fraternity basketball team is 6' 4").

But before we start let us introduce the two foundations of statistical theory; the **Normal Distribution**, and the **Central Limit Theorem**.

## THE NORMAL OR GAUSSIAN DISTRIBUTION

If we measure:

- the height of all the male students at our university
- the IQ test results of all Oregon third graders.
- the physical and mental characteristics (head size, brain weight, cleverness etc.) of a racial group of people.
- 100 000 independent measurements of the length of a piece of wood
- the daily reading of the octane number of gasoline produced from a refinery
- the horizontal scatter of bullets on a target

we find, in these and in many other situations, that the so-called **Normal** or **Gaussian distribution curve** results. This is a special bell shaped curve represented by the equation:

$$y = \frac{1}{\sigma\sqrt{2\pi}} e^{\frac{-(x-\mu)^2}{2\sigma^2}}$$

where $x$ represents the individual readings of the data and $y$ the fraction of readings at that x value.

The unusual feature of this curve is that just two numbers, the <u>mean of the population</u> $\mu$ and the <u>standard deviation of the population</u> $\sigma$, or more usefully, its square, called the <u>variance</u> $\sigma^2$, will describe and completely define the curve. We illustrate these two quantities in the sketches on the following page:

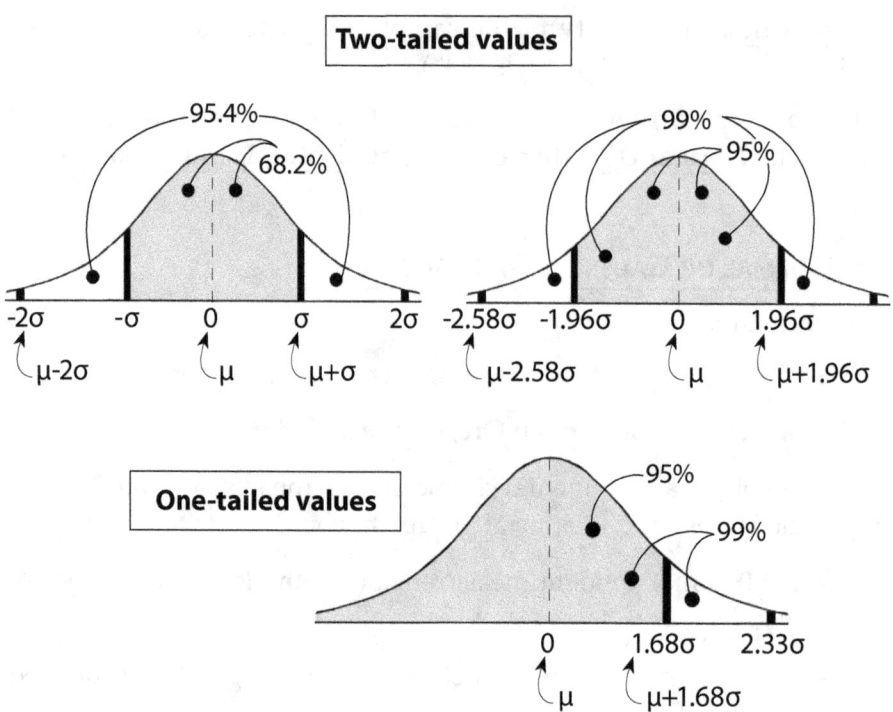

How do we measure or find $\mu$ and $\sigma^2$?

$$\mu = \frac{\Sigma x}{W} \text{....where } W \text{ is the size of the population} \quad (1)$$

$$\sigma^2 = \frac{\Sigma(x-\mu)^2}{W} = \frac{\Sigma(x^2) - \frac{(\Sigma x)^2}{W}}{W} \quad (2)$$

⟵ useful computing formula

**Basic assumption:** In the statistical tests that we will present we always assume that the population of readings approximates the Normal distribution. If it does not, then we are stuck and do not know how to analyze the situation. So what do we do? We shrug our shoulders, assume the Normal distribution and hope that we won't be far wrong.

## THE CENTRAL LIMIT THEOREM

Consider a population of W observations with mean $\mu$ and variance $\sigma^2$. If we draw samples, <u>all possible samples of size n</u> from this population what can we say about these samples, each having n readings, and their means $\bar{x}$

1. The mean of all the sample means, $\bar{x}$, is equal to the population mean ($\bar{\bar{x}}$) = $\mu$
2. The distribution of sample means, $\bar{x}$, is Normal with variance $\sigma^2/n$

The larger $n$ becomes the more concentrated about $\mu$ does the distribution curve of $\bar{x}$ become. These two findings represent the Central Limit Theorem.

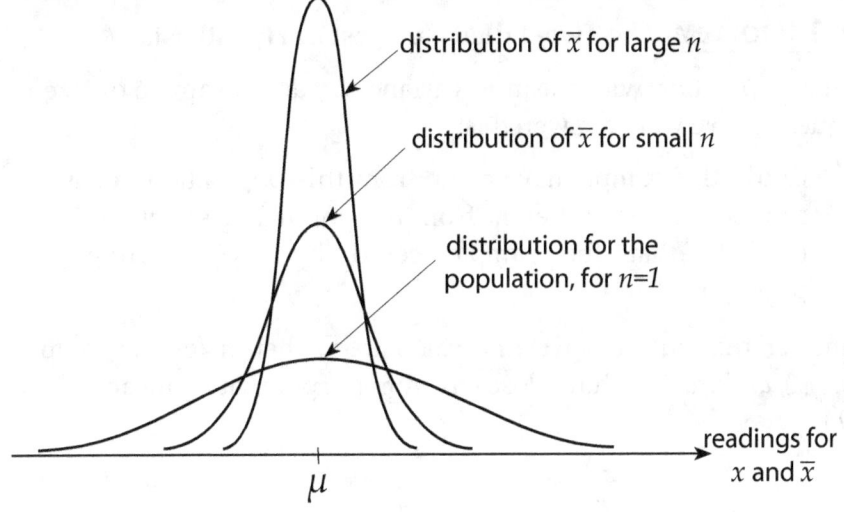

## THE THREE TYPES OF STATISTICAL TESTS

We now have enough background information (the Normal distribution, and the Central Limit Theorem) to start making statistical inferences. These are usually done in three ways.

**Type 1 Problem.** Given a population with mean $\mu$ and variance $\sigma^2$, known, plus a sample of readings $x_1, x_2, ... x_n$ we ask whether the sample could reasonably have come from the population. This is called the <u>test of hypothesis</u> because we start with the hypothesis (or guess) that the sample does in fact come from the population, and then see whether we are justified in doing so.

**Type 2 Problem.** Same as Type 1 problem except that we do not know the variance of the population.

**Type 3 Problem.** Here we ask from what kind of population could we have (reasonably) pulled this sample. This test is called the estimation of a population parameter. Here we are not comparing a sample with a population - we are only dealing with a sample, and guessing what kind of population it could have come from.

In all three cases we must state how much faith we have in our conclusions.

### TYPE 1 PROBLEM  The Two-Tailed "u" Test of Hypothesis

Given a population with mean $\mu$, variance $\sigma^2$, and a sample of size $n$. The question we want answered is:

Could this sample have come from this population, or is the sample mean so different from the population's that we feel it unlikely that the sample could have come from this population?

To answer this question we first calculate $\bar{x}$, then $u$ for the sample, (or $u_{sample}$), and then see how close it is from the population mean (where $u = 0$)

$$\text{where} \quad \bar{x} = \frac{\sum x}{n} \quad \text{and} \quad u_{sample} = \frac{\bar{x} - \mu}{\frac{\sigma}{\sqrt{n}}} \quad (3)$$

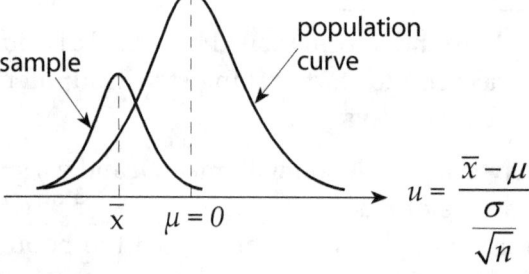

$$u = \frac{\bar{x} - \mu}{\frac{\sigma}{\sqrt{n}}}$$

**The Significance Level**  This is the probability of rejecting the correct hypothesis that the sample could have come from the population. This figure is often set at 5% ( or 1% ).

If the sample mean is close to the population mean (or $u_{sample}$ is close to zero) we then conclude that the sample could have come from the population. However if the data values are far from the population mean (or $u_{sample}$ is far from zero) then we conclude that the sample likely could come from another population. So here are the critical values for $u$

Table

| significance level, | 10% | 5% | 2% | 1% | 0.1% |
|---|---|---|---|---|---|
| $u_{critical} = u_{table}$ | 1.645 | 1.960 | 2.326 | 2.576 | 3.291 |

What this means is that if $|u_{sample}| > u_{table}$ then we decide that the sample does not come from the population. If $|u_{sample}| < u_{table}$ then there is no reason to doubt that the sample comes from the population.

## TYPE 2 PROBLEM  The Two-Tailed "t" Test of Hypothesis

Here we do not know the variance of the population in which case the distribution of all possible $x$ values is bell shaped about $\mu$, a bit broader than Normal and dependent on the sample size, $n$. We show this below

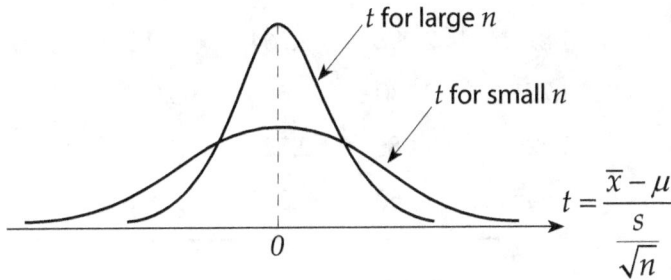

$$t = \frac{\bar{x} - \mu}{\frac{s}{\sqrt{n}}}$$

We call this the "t" distribution. Now calculate the $t$ value from our data

$$t_{sample} = \frac{\bar{x} - \mu}{\frac{s}{\sqrt{n}}} \qquad (4)$$

where

$$s^2 = \frac{\Sigma(x - \bar{x})^2}{n-1} = \frac{\Sigma x^2 - \frac{(\Sigma x)^2}{n}}{n-1} \qquad (5)$$

The "t" values at various significance levels and sample sizes are presented in practically every statistics book. Here is an abbreviated table of values:

**The t Table**

Significance level - two-tailed test

| Degrees of Freedom, D.F. = n - 1 | 10% | 5% | 2% | 1% |
|---|---|---|---|---|
| 1 | 6.314 | 12.706 | 31.821 | 63.657 |
| 2 | 2.920 | 4.303 | 6.965 | 9.925 |
| 3 | 2.353 | 3.182 | 4.541 | 5.841 |
| 4 | 2.132 | 2.776 | 3.757 | 4.604 |
| 5 | 2.015 | 2.571 | 3.365 | 4.032 |
| 7 | 1.895 | 2.365 | 2.998 | 3.499 |
| 10 | 1.812 | 2.228 | 2.764 | 3.169 |
| 13 | 1.771 | 2.130 | 2.650 | 3.012 |
| 20 | 1.725 | 2.086 | 2.528 | 2.845 |
| $\infty$ | 1.645 | 1.960 | 2.326 | 2.576 |
| | 5% | 2.5% | 1% | 0.5% |

t values

Significance level - one-tailed test

### Conclusion

For the significance level chosen, say 5%, to test the hypothesis we have to see whether the $t$ value from our sample lies within the interval of $+t$ to $-t$ for the significance level and sample size, or outside the interval. If $t$ sample lies within the interval then there is no good reason to doubt that the sample comes from the population.

If $t_{sample}$ lies outside the interval of $-t$ to $+t$ then we reject the hypothesis. This means that the chance that this sample could have come from the population is very unlikely and less than 5%.

## OUTCOMES

We have four possible outcomes, as shown in the following table:

|  | If we conclude that the data does come from the population (if we accept the hypothesis) | If we conclude that the data does not come from the population (if we reject the hypothesis) |
|---|---|---|
| If the data actually comes from the population (If the hypothesis is correct) | No error | type I error<br>we made a mistake in not hiring him |
| If the data does not actually come from the population (If the hypothesis is not correct) | type II error<br>not enough evidence to reject | No error |

Set by the significance level, 5%, 1% . . .

A pressing problem is how to balance these two errors because as Type I error increases Type II error decreases and vice versa.

As illustration design a test to help us choose who to hire into our firm. Here are possible results. We start with the hypothesis that the test is designed to find a good man.

|  | Passes test. Hire him | Fails test. Reject him |
|---|---|---|
| Hypothesis is correct; he is actually a good man | we are happy | we made a mistake in not hiring him (type I error) |
| Hypothesis is incorrect; he'd be a disaster | he was lucky, or we made a mistake (type II error) | we were correct in not hiring him |

If the test is too hard we will reject too many potentially good prospects (type I error). If it is too easy we hire too many incompetents (type II error).

**One- and two- tailed tests.** So far we have tested whether a sample has fallen in the interval between $-t$ and $+t$, or whether it has fallen outside and to the left or right of the interval. This is called a two-tailed test. Thus at the 5% significance level:

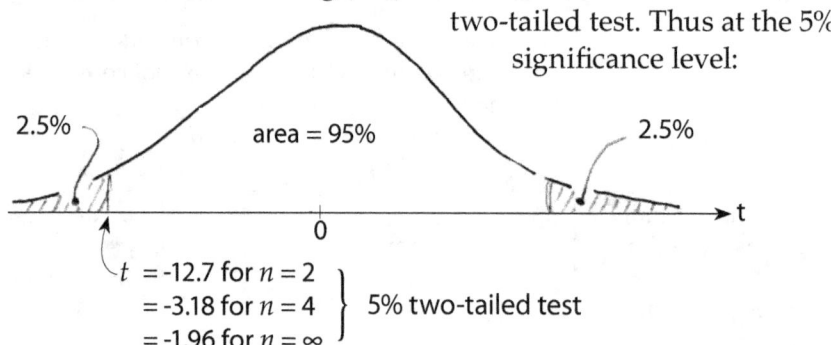

There are some situations where we are only concerned with whether the sample falls on one side of the mean. For example, suppose we want to know whether the butcher is short-weighing the 16 oz. packages of hamburgers. We don't care if he gives us more than 16 oz. To answer such questions we want to use a one-tailed test for $\mu=16$ oz. At the 5% significance limit for a sample of $n = 11$ readings this is shown in the sketch below.

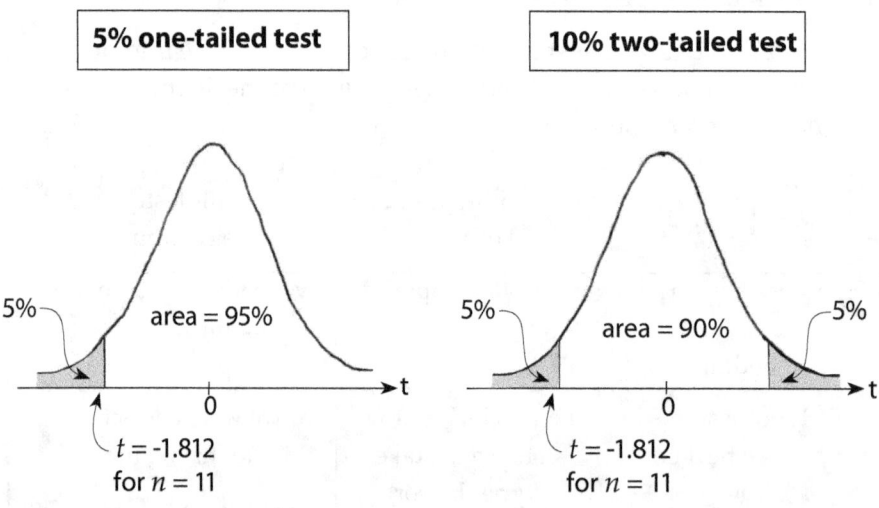

## TYPE 2 PROBLEM

**EXAMPLE OF A TYPE 2 PROBLEM** One-tailed "t" Test of Hypothesis

Every Thursday my butcher has a one pound package of hamburger wrapped and ready for me. He is a pleasant chap and we are on friendly terms. Recently I bought some scales and out of curiosity I weighed the hamburger packages. The results made me even more curious so I kept a record of the weights. This is what I found (in ounces) for the next nine packages sold to me.

$$13, 17, 16, 16, 15, 13, 14, 15, 16$$

Now, I wonder if my friendly neighborhood butcher has been cheating me? But before I switch to another butcher or before I have a row with him, I'd like to be at least 95% sure that that I am getting 16 oz. or more in each package, on the average.

**Solution:** I am only interested in whether I am getting 16 oz or more, so I will use a one-tailed test. Thus the hypothesis to be tested is $\mu \geq 16$ oz.

The alternative hypothesis: $\mu < 16$ oz
The significance level will be 5%, one tailed test
The number of data points: n = 9, or D.F.= 8

$$|t_{critical,\ n=9}| = \sim 1.86, \quad \text{from the t - table}$$

$$t_{sample} = \frac{\bar{x} - \mu}{\frac{s}{\sqrt{n}}}$$

$$\bar{x} = \frac{(13+17+16+16+15+13+14+15+16)}{9} = 15 \text{ oz}$$

$$s^2 = \frac{(13-15)^2 + (17-15)^2 + \ldots}{9-1} = \frac{16}{8} = 2$$

$$t_{sample} = \frac{|15-16|}{\frac{\sqrt{2}}{\sqrt{9}}} = 2.12, \text{ and this is greater than } t_{critical}$$

## Conclusion:

Compare $t_{sample}$ with $t_{critical}$. At the 5% one-tailed significance level I reject the hypothesis that $\mu \geq 16$ oz.

Interpretation: Either my scales are not accurate or I'm 95% sure that the butcher is cheating me; and since I do not like to argue, I'll switch butchers.

**TYPE 3 PROBLEM** Estimating the Location of the Population Mean (variance of the population is also unknown).

Given a sample of size $n$ with mean $\bar{x}$, estimate where $\mu$ could be. Certainly the best estimate for $\mu$ is to say that $\mu = \bar{x}$. However there is no measure of reliability for such an estimate. Also the estimated number for $\mu$ would certainly not be exact.

Alternatively we could specify an interval for $\mu$, the <u>confidence interval</u>, and assign a number to tell the confidence we have that $\mu$ is likely to be in that interval. We call this the <u>confidence coefficient</u>, often chosen at 95% or 99%.

### Procedure

**Step 1.** State the problem. Estimate $\mu$ from the evidence of the sample of size $n$.

**Step 2.** Calculate the two statistics below. From eqs, 4 and 5

$$\bar{x} = \frac{\Sigma x}{n} \quad \text{and} \quad s^2 = \frac{\Sigma(x-\bar{x})^2}{(n-1)} = \frac{\Sigma(x^2) - \frac{(\Sigma x)^2}{n}}{(n-1)}$$

**Step 3.** Now $\mu$ lies close to $\bar{x}$ with a spread between them given by the $t$ distribution. This is shown below at the 95% confidence coefficient which is equivalent to the 5% significance level for $n$ or for the degrees of freedom $= n - 1$

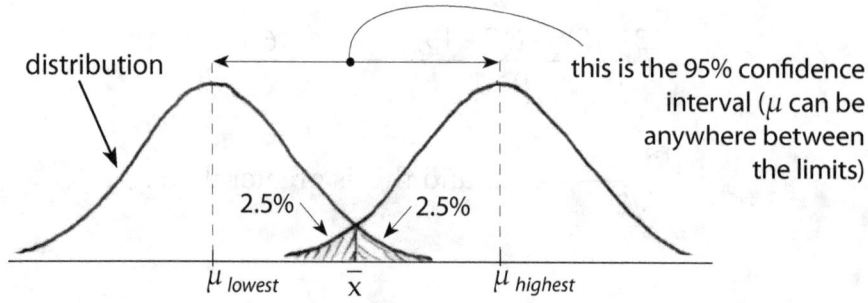

# TYPE 3 EXAMPLE

Therefore the confidence limits for $\mu$ are $\mu = \bar{x} \pm t \cdot \dfrac{s}{\sqrt{n}}$

Thus the chances are 95% that the population mean is in the interval between $\mu_{lowest}$ and $\mu_{highest}$.

## EXAMPLE OF A TYPE 3 PROBLEM  Estimating the Location of a Population Mean

To the nearest 5 lbs., the sophomore ChE students at our Univ. weigh:

145, 230, 160, 200, 130, 130, 190, 175, 175, 140, 170, 130, 125, 165 pounds

Would you please estimate the mean weight of all sophomore students in the country from this sample, assuming that our students represent a random sample. Let the reliability of your estimate be 95%.

**Solution:** This is a type 3 test where we are estimating the location of a population parameter. Estimate the weight by using a confidence interval where the confidence coefficient is 95%

The number of data points is: n = 14

So the degrees of freedom, D.F. = $n - 1 = 13$

Thus from pg. 7.8: $t_{95\% \text{ two tailed}} = t_{crit} = 2.130$

Calculate $\bar{x} \cong \dfrac{(145 + 230 \ldots)}{14} \cong 162$

Calculate $s^2 = \dfrac{\Sigma(x-\bar{x})^2}{n-1} = \dfrac{17^2 + 16^2 + \ldots}{13} = \dfrac{12481}{13} = 960$

Calculate $s = \sqrt{960} = 31$ pounds

So the confidence limits for the population mean is

$\mu = \bar{x} \pm (t_{crit}) \cdot \dfrac{s}{\sqrt{n}} = 162 \pm (2.13) \cdot \dfrac{31}{\sqrt{14}} = 162 \pm 17.6$

Upper limit = 180 lbs
Lower limit = 144 lbs

In everyday language this means that the average weight of all Chemical Engineering sophomores in the US should be between 144 and 180 pounds, and I am 95% sure of that conclusion.

## RELATIONSHIP BETWEEN Y AND X

In his curiosity, in his quest for knowledge, or in the development of science, man wants to know how this is related to, affects or causes that. For example:

- how does aspirin affect headaches
- how much does the victim suffer when cursed by the witch doctor,
- how does a rise in temperature lengthen a rod of metal
- how does exercising improve a person's health

In general we want to know how $y$ is related to $x$. We can show this on an $x$ vs $y$ plot.

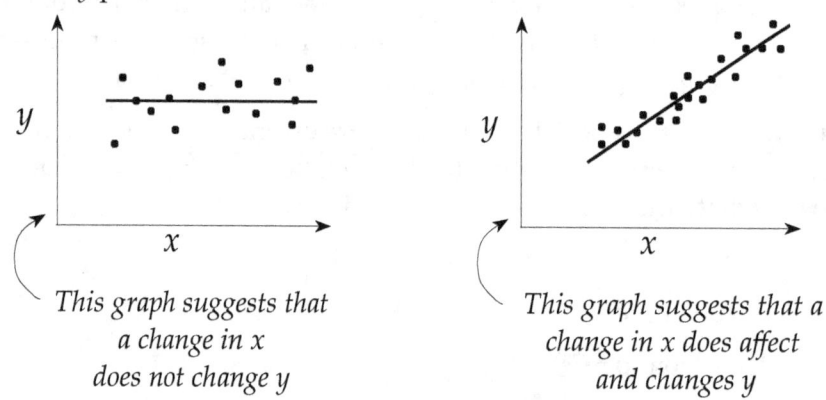

This graph suggests that a change in $x$ does not change $y$

This graph suggests that a change in $x$ does affect and changes $y$

When we take some $x$ and $y$ data which has some scatter of $y$ values about $x$ we want to know how sure we are that $x$ and $y$ are related - strongly, likely, probably, not at all, etc. - and what is the relationship. To answer these questions we turn to statistics for help.

We have to answer some preliminary questions before we proceed with an analysis.

1. Does a straight line fit the data well, or does a curve fit the data better?

2. Is the spread of the data the same at all $x$, or does it differ?

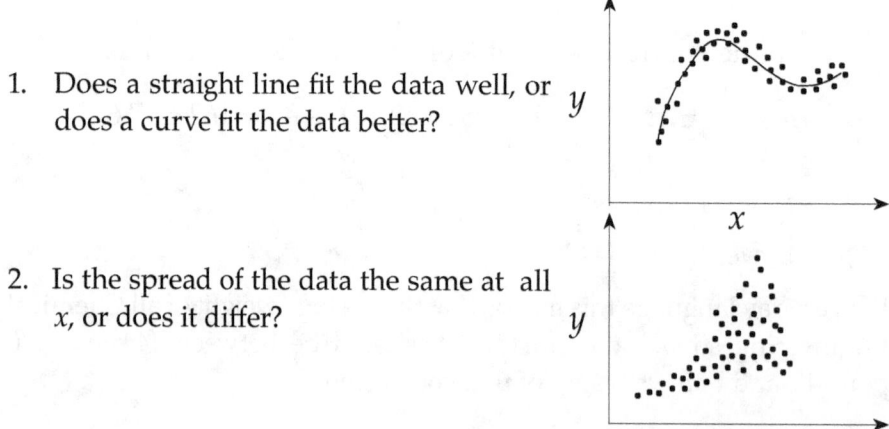

# Linear Relationship and the Best Line through some Data

The answers to these questions suggests how we should treat the x - y data. So let us consider the simplest situation, and describe the basic idea of statistics.

## LINEAR RELATIONSHIP AND THE BEST LINE THROUGH SOME DATA

Consider a theoretical line or prediction represented by $Y = A + Bx$, a sample of n data points $x_1y_1, x_2y_2, \ldots, x_ny_n$

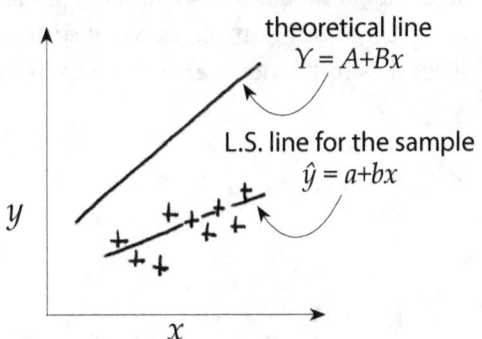

and before we start we must accept

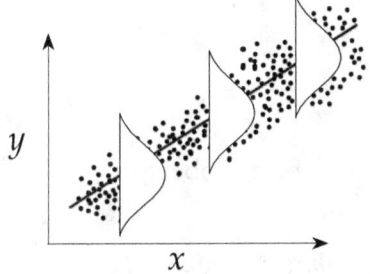

- that the theory has a linear relationship of $y$ with $x$,
- that $x$ is well known,
- and that the spread of $y$ is the same at all $x$

In general we want to know one of a number of things: whether the sample could have come from the theoretical prediction, or have the same slope or intercept as the prediction. In this chapter we just see whether the sample has the same slope as the prediction. To do this we proceed with a three step procedure, as follows:

**Step 1** Find the best line to represent the data. This line is called the least square line (LS line), $\hat{y} = a + bx$

**Step 2** Find the scatter of the data about its LS line.

**Step 3** Make the statistical test to tell whether this sample could reasonably have the same slope as the theoretical line, $Y = A + Bx$: or whether it is unlikely to have. This is another version of the test of hypothesis.

In this procedure we start by assuming that the sample does in fact have the same slope as theory (this is called the NULL hypothesis). If the slope of our sample is very different from the theoretical then we reject this hypothesis and have to use an ALTERNATIVE hypothesis.

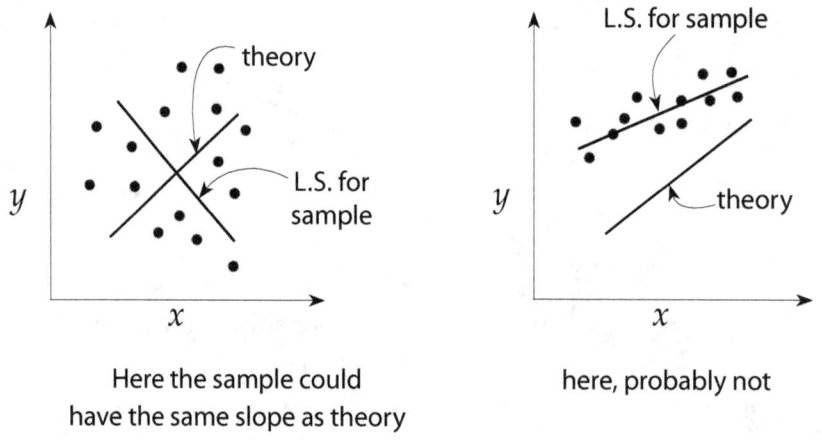

Here the sample could have the same slope as theory

here, probably not

*The null hypothesis says that the sample has the same slope as theory*

The SIGNIFICANCE LEVEL tells how unusual our sample has to be before we say that we do not believe our null hypothesis. In hypothesis testing this cut off point (when you accept or reject) is often chosen at 5%, 1%, or 0.1%. This means that if our null hypothesis would produce this kind of data less than 5%, 1% or 0.1% of the time then we say we do not believe that our null hypothesis could be true.

The test that is used here is the same test as used in Chapter 6, the t-test, and it is done in 3 steps.

**Step 1** Finding the Least Squares Line (the LS line) for the Sample

# FIND THE L.S. LINE

## 7.17

This procedure shows how to get the proper, or best line, to represent the data. So for the n points calculate:

$$\Sigma x = x_1 + x_2 + ... + x_n$$

$$\bar{x} = \frac{\Sigma x}{n}$$

$$\Sigma y = y_1 + y_2 + ... + y_n$$

$$\bar{y} = \frac{\Sigma y}{n}$$

$$\Sigma x^2 = x_1^2 + x_2^2 + ... + x_n^2$$

$$\Sigma y^2 = y_1^2 + y_2^2 + ... + y_n^2$$

$$\Sigma xy = x_1 y_1 + x_2 y_2 + ... + x_n y_n$$

$$\Sigma' x^2 = \Sigma(x - \bar{x})^2 = \Sigma x^2 - \frac{(\Sigma x)^2}{n}$$

$$\Sigma' y^2 = \Sigma(y - \bar{y})^2 = \Sigma y^2 - \frac{(\Sigma y)^2}{n}$$

$$\Sigma' xy = \Sigma(x - \bar{x})(y - \bar{y}) = \Sigma xy - \frac{(\Sigma x)(\Sigma y)}{n}$$

Here $\Sigma'$ is an abbreviation of what follows

$$\boxed{b = \frac{\Sigma' xy}{\Sigma' x^2}} \quad \boxed{a = \bar{y} - b\bar{x}}$$

And the LS line is: $\boxed{\hat{y} = a + bx}$

where $\hat{y}$ is the LS value (best estimate) of $y$

## Step 2 Measure of Scatter of Data Points from its LS line

We first introduce the concept of sum of squares, SS, needed in the calculation

<u>Total SS</u> This is the total sum of squares of the deviation of the data points from the mean value of $y$, or $\bar{y}$

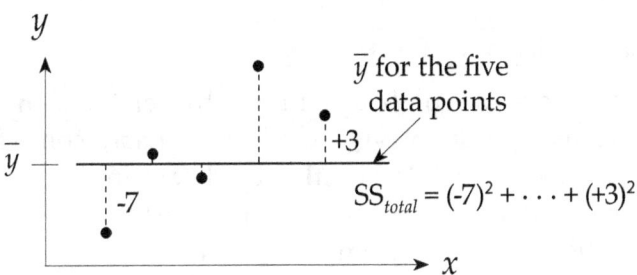

$\bar{y}$ for the five data points

$SS_{total} = (-7)^2 + \cdots + (+3)^2$

## SS of Data from the LS Line

The least squares line gives the smallest remaining SS for the n data points.

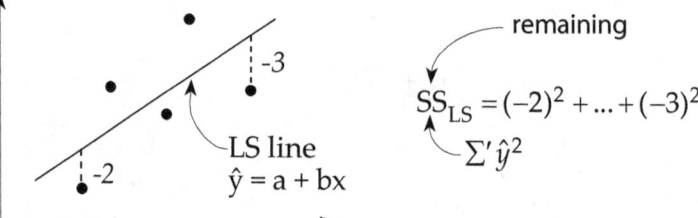

SS Removed by the LS line  $SS_{removed} = SS_{total} - SS_{LS}$

## Other Useful Relations

After a lot of tedious mathematics we find:

$$\underbrace{\Sigma' \hat{y}^2}_{\substack{\text{SS remaining}\\\text{after using}\\\text{the LS line}}} = \Sigma(y - \hat{y})^2 = \underbrace{\Sigma' y^2}_{\text{total SS}} - \underbrace{b\Sigma' xy}_{\substack{\text{SS removed}\\\text{by the LS}\\\text{line}}}$$

$$\boxed{s^2(\hat{y})} = \frac{\Sigma' \hat{y}^2}{n-2}$$

$$\boxed{s^2(b)} = \frac{s^2(\hat{y})}{\Sigma' x^2} = \frac{\Sigma' \hat{y}^2}{(n-2)\Sigma' x^2}$$

Correlation coefficient = r

and $\underbrace{r^2}_{\substack{\text{fraction of SS}\\\text{removed by LS line}}} = b \frac{\Sigma' xy}{\Sigma' y^2} = \frac{\text{SS removed by LS line}}{\underbrace{\text{original SS}}_{\text{total SS}}}$

### Step 3  Calculation for the t-test

Here is how we go about telling whether the sample ($\hat{y} = a+bx$, with scatter measured by $s(y)$) could reasonably come from our theory, $Y = A + Bx$, or not. We do this by determining the "t" value for our sample of data (we show how to calculate this below) and compare it to the critical value "$t_{crit}$" from the tables in this chapter for the significance level we have chosen. Then . . .

# THE t-TEST

if $|t| < t_{crit}$ there is no reason to reject our hypothesis, our null hypothesis. This means that there is no reason to doubt that the sample has the same slope as the theoretical line.

if $|t| > t_{crit}$ the result is very unusual (say less than say 1% of the time) so the sample is not related to the population and we reject the hypothesis and will have to look for a different explanation.

This type of test represents what is called <u>Regression Analysis</u> and here are a number of such tests

**Test A** Could our data have come from a theoretical line of zero slope, or is the slope non zero?

**Hyp: b = 0**

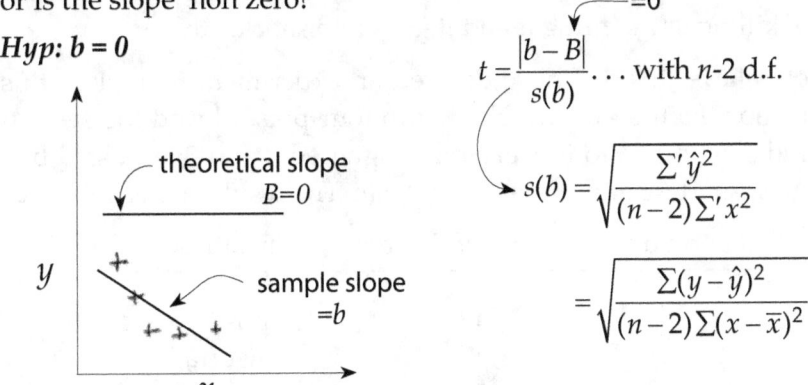

$$t = \frac{|b - B|}{s(b)} \quad \text{... with } n\text{-2 d.f.}$$

$$s(b) = \sqrt{\frac{\sum' \hat{y}^2}{(n-2)\sum' x^2}}$$

$$= \sqrt{\frac{\sum(y - \hat{y})^2}{(n-2)\sum(x - \bar{x})^2}}$$

**Test B** Could our data come from a theoretical line of slope B, or is the slope likely different from B?

**Hyp: b = B**

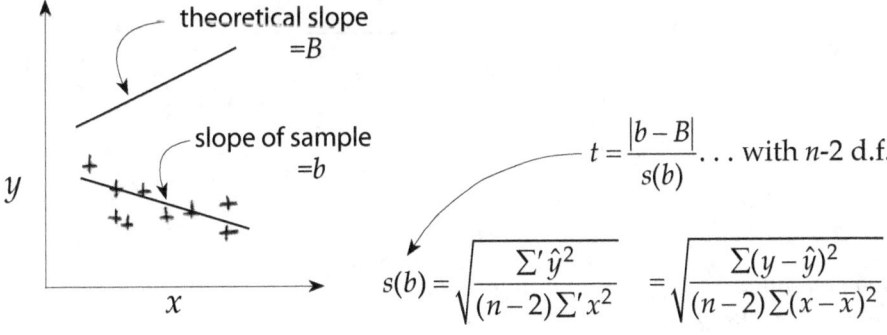

$$t = \frac{|b - B|}{s(b)} \quad \text{... with } n\text{-2 d.f.}$$

$$s(b) = \sqrt{\frac{\sum' \hat{y}^2}{(n-2)\sum' x^2}} \quad = \sqrt{\frac{\sum(y - \hat{y})^2}{(n-2)\sum(x - \bar{x})^2}}$$

## Illustrative Example 1 • Does Music Help Plants Grow?

Throughout time man had suspected that plants responded to our feelings. If we are sympathetic, if we love them, and if we think nice thoughts about them they will grow happily and vigorously -- otherwise they won't. This feeling can be found in all the religions of agrarian societies, witness the Spring planting festivals, etc.

Did you know that Charles Darwin played his tuba to his flowers to see if they would show any response?

Recently "Psychology Today" reports that the electrical waves of a plant responded to our thoughts about the plant. Just walk up to a plant and think of killing or maiming it and it goes into shock, etc.

I do not believe this so I have devised an experiment to explore this phenomenon. Four seeds are planted in four pots. I loved the seed in Pot 1 and sang songs to it every day. I ignored 2 and 3; and said bad things to 4, snarling and threatening it, etc. The results are as follows:

| plant and pot | number of flowers | my attitude |
|---|---|---|
| 1 | 19 | positive |
| 2 | 17 | neutral |
| 3 | 15 | neutral |
| 4 | 13 | negative |

Let me put this data in the form below

| my attitude ($x$) | plants' response ($y$) |
|---|---|
| 1 | 3 |
| 0 | 2 |
| 0 | 1 |
| -1 | 0 |

*Do you think that there is anything to this suspicion that plants respond ($y$) to what people think ($x$)?*

# EXAMPLE 7.21

**Solution**

Plot the four $x$–$y$ data points and assume that they are fitted by a straight line

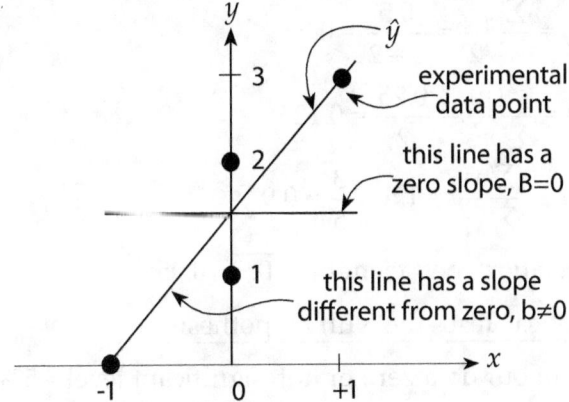

We want to know whether the slope of the line really differs from zero. For this compare $t_{crit}$ with $t_{sample}$

From the above table, the 5% value for $n = 4$ or d.f. $= 2$ we have $t_{crit} = 4.303$. From our data calculate:

$$t = \frac{|b - B|}{s(b)}$$

**Step 1** First find the LS line to represent the data

$\Sigma x = 0$  $\quad \Sigma x^2 = 2 \quad\quad \Sigma' x^2 = \Sigma x^2 - \frac{(\Sigma x)^2}{n} = 2 - 0 = 2$

$\Sigma y = 6$  $\quad \Sigma y^2 = 14 \quad\quad \Sigma' y^2 = \Sigma y^2 - \frac{(\Sigma y)^2}{n} = 14 - 9 = 5$

$\bar{x} = 0$

$\quad\quad\quad\quad \Sigma xy = 3 \quad\quad \Sigma' xy = \Sigma xy - \frac{(\Sigma x)(\Sigma y)}{n} = 3 - 0 = 3$

$\bar{y} = 1.5$

Therefore  $b = \dfrac{\Sigma' xy}{\Sigma' x^2} = \dfrac{3}{2} = 1.5$

$a = \bar{y} - b\bar{x} = 1.5 - 1.5\,(0) = 1.5$

So the LS line is  $\hat{y} = 1.5 + 1.5x$

**Step 2** Measure the scatter of the data

$$\Sigma'\hat{y}^2 = \Sigma'y^2 - b \cdot \Sigma'xy = 5 - 4.5 = 0.5$$

$$s^2(\hat{y}) = \frac{\Sigma'\hat{y}^2}{n-2} = \frac{0.5}{4-2} = 0.25$$

$$s^2(b) = \frac{s^2(\hat{y})}{\Sigma'x^2} = \frac{0.25}{2} = 0.125$$

$$r^2 = b \cdot \frac{\Sigma'xy}{\Sigma'y^2} = 1.5 \cdot \frac{3}{5} = 0.9$$

Correlation coefficient $r = \sqrt{0.9} = 0.95$

**Step 3** Statistical Test - the Null Hypothesis

—Is the slope of our data zero or not, significant level = 5%?

Compare $t_{crit}$ with $t$

$$t_{crit} = 4.303$$

$$t = \frac{|b-B|}{s(b)} = \frac{1.5-0}{\sqrt{1.25}} = 1.34$$

Since $t < t_{crit}$, we come to the following conclusions:

- we have no reason to suspect that playing music has any effect on plant growth

- There is not enough evidence showing that playing music affects plant growth positively or negatively.

If we took more data we'd have a "sharper" test, which means having a better chance to tell whether music affects or does not affect plant growth.

## Tables of Critical "t" values for y vs x Problem

| | Signicance level - two tailed test | | | |
|---|---|---|---|---|
| D.F. = n - 2 | 10% | 5% | 2% | 1% |
| 1 | 6.314 | 12.706 | 31.821 | 63.657 |
| 2 | 2.920 | 4.303 | 6.965 | 9.925 |
| 3 | 2.353 | 3.182 | 4.541 | 5.841 |
| 4 | 2.132 | 2.776 | 3.757 | 4.604 |
| 5 | 2.015 | 2.571 | 3.365 | 4.032 |
| 7 | 1.895 | 2.365 | 2.998 | 3.499 |
| 10 | 1.812 | 2.228 | 2.764 | 3.169 |
| ∞ | 1.645 | 1.960 | 2.326 | 2.576 |
| | 5% | 2.5% | 1% | 0.5% |

Significance level - one tailed test

## AFTERTHOUGHTS

The Relationship between Mathematics and Statistics

It is interesting that mathematics has a very ancient history and today is taught to all children in grammar school and also in colleges. For example Pythagoras' School came up with the Pythagorean Theorem about 570 B.C. However it was known in Mesopotamia long before then, also in India and other ancient societies. Today we have over 370 different proofs of this theorem, one of them created by a US President, James Garfield.

As opposed to the deductive activity of mathematics, statistics deals with inductive inferences by saying that something is likely to be so about a population by looking at a sample.

This qualitative process started, not thousands of years ago, but just about 100 years ago, in the first decade of the twentieth century. The Student t-test was the trigger for this development. It is the prime tool of the applied scholar, the engineer, the technologist, the agriculturist, the biologist, the life scientist, etc. Here we say that something is likely to be so of a population with a certain probability from data gotten from a sample.

### W. S. Gossett, the Isaac Newton of Statistics

William Sealey Gossett formulated the t-test in the early 1900 while he worked as a chemist for Guiness Brewery. It was used to judge whether a new batch of beer matched a standard in taste.

This test became a company secret which then prevented Gossett from trying to publish the method. Finally he was allowed to publish it under the fictitious pen name of "Student". Today this is called the Student t-test. The Student t-test has been extended to the Fisher F-test.

Gossett's seminal development opened the flood gates to the **study of statistics**

## PROBLEMS

*The race is not always to the swift, nor the battle to the strong, but that's the way to bet.*

1. Under what conditions is the sample mean Normally distributed?

2. In testing a hypothesis at a given significance level why do we like to have a large sample?

3. If the observations of the population are in units of $\$.hr.m^3$ what are the units of
    a) $\bar{x}$
    b) $\sigma^{2(x)}$
    c) $t$
    d) $s(\bar{x})$

4. In testing a hypothesis with the same significance level, will it be more likely to reject a true hypothesis with a large or small sample?

5. In testing a hypothesis with the same significance level will it be more likely to reject a false hypothesis with a large or small sample?

6. Do the lengths of the confidence intervals of a population mean vary from sample to sample of the same size when using the same confidence coefficient? Explain.

7. In testing the hypothesis that a sample comes from a population we may use a large or a small sample. In which case will the hypothesis be more likely to be accepted? Why?

8. Is a bell shaped curve always a Normal curve?

9. If a hypothesis is already rejected by a sample of 10 observations, is it likely to be rejected by a sample of 100 observations? Why?

10. The 95% confidence interval of the population mean as obtained from a given sample is 49 to 51. Is it correct to say

that 95 out of 100 times, the population mean falls inside the this interval? Whether your answer is yes or no, state your reason.

11. In the test of hypothesis what is the consequence of reducing the significance level from 5% to 1% without changing the sample size?

12. Under what conditions does the t-distribution become a Normal distribution?

13. What is the advantage of having a large sample in determining the confidence interval of a population mean?

14. What is the consequence of using a zero percent significance level?

15. Does one need a large or small sample to reject a false hypothesis which is very close to the true one?

16. We use the Normal distribution so frequently in statistics class. What is an abnormal distribution ?

17. How do you divide 24 children into two groups at random?

18. What is the consequence of using a 99% confidence coefficient instead of a 95% confidence coefficient in the estimation of a population parameter?

## Qualitative Questions

In the following two problems: pb 19 and pb 20
a) What is the hypothesis that you will test?
b) What is the alternative hypothesis?
c) If you accept the hypothesis what does it mean in everyday words?
d) If you reject the hypothesis what does it mean in everyday words?
e) Would you use a one-tailed or two-tailed test?

19. Old time Syrian camel drivers swear that a pinch of female camel dung added just before fermentation is essential for producing a yogurt of the highest quality and most

delectable and enrapturing flavor. I doubt it, so I have prepared two batches of product, one with and one without this magic ingredient. I've asked yogurt aficionados to taste-test samples of both and give me their conclusions at the 5% significance level.

20. So you claim that Bufferin relieves headaches faster than does aspirin? Well I've made a survey and I'm ready to analyze the data statistically at the 1% significance level to check your claim, to see if it is true or whether it represents false advertising.

21. What is your confidence coefficient? Let's see. You go to the little coffee house around the corner for your 'pick-me-up' at 11 am every day. Suppose that one day the person next to you in line suggests flipping a coin to see who pays for today's coffee. You agree - and you lose. It's no big deal.

    Next day you meet the same person in line and you lose again. Although you do not suspect that there is any funny business involved, still how many times would you play with that person before you decide to avoid him. This fellow:
    a) is pleasant but is a stranger,
    b) is your statistics professor,
    c) is the minister of your church.
    From this decision of yours can you determine your confidence coefficient?

## QUANTITATIVE PROBLEMS

22. By means of both the definition formula and the computing formula, calculate the standard deviation and the variance of the following population: 8, 9, 5, 8, 5.

23. Consider the following observations as a sample: 8, 3, 4. Find the unbiased estimate of the population mean and of the population variance.

24. A sample mean was found to be 40.01. Do you accept the hypothesis that the population mean is 40. A mere yes or no is not a sufficient answer.

25. The mean and standard deviation of a set of 500 observations are 50 and 10 respectively. What is the new mean and new standard deviation if
   a) 9 is subtracted from each observation.
   b) each observation is divided by 5.
   c) 50 is subtracted from each observation and the result is divided by 10.

26. The sample mean based on 10,000 observations is 23.12 and $s$ is 1. Is the population mean equal to 23.00? State how you reached your conclusion.

27. A sample consisting of the observations 2, 3, 5, 2 was drawn from a population whose variance is known to be 25. Using the 1% significance level test the hypothesis that the population mean is equal to 4. In working this problem, give the hypothesis, alternatives to this hypothesis, assumptions, level of significance, critical regions, results of computation and conclusion.

28. Theory A says that the average weight of all adult kangaroos should be 50 kg, while theory B predicts that they should weigh 75 kg. You wish to choose between these two theories so off you go to weigh kangaroos. You manage to catch 6 which weigh 55, 56, 54, 55, 57, 53. What do you conclude? Use the method for the estimation of a population parameter.

29. What is the mean of the *t*-distribution?

30. The size of a sample is 25 and the variance for the distribution of all sample means of that size, is 2.4. How large does the size of the sample have to be if we want to reduce to it 1.5.

31. The distribution of height of the U.S. adult male population follows a Normal distribution with a mean of 68" and a standard deviation of 2". Mr. X is 71" tall. What can you conclude?

32. Suppose that the mean lifetime of a certain type of electronic tube is 10,000 hrs with a standard deviation of 800 hrs. The engineers now develop a new design of tube. A sample of 64 of the new-type tubes is tested, and the mean life of this sample is found to be 10,200 hrs.

    Is the apparent improvement of 200 real or could it be due to chance variation?

33. The NIH report based on tests on many patients tells that 30 days on the standard cholesterol lowering drug 'Provachol' lowers one's cholesterol count by 25 points. But I think that our new drug called 'No-Cool' can do better. We tested it on 5 patients and found the following drop in cholesterol count of: 45, 20, 55, 5, 75 points. Can we claim with 95% certainty that our drug is superior to Provachol?

34. <u>Newton versus Einstein.</u> Whenever two different theories, hypotheses or models are proposed to represent actual physical phenomena, scientists must decide which is the better of the two. To do this they draw deductions from the two competing theories until one theory says "this should be the case" and the other says "that should be the case." Then an experiment is planned to determine what actually happens to tell which theory wins the day. This is called the "crucial experiment," and it always is the high point in science, especially so when the theories concerned are very important.

    This situation occurred recently, in 1915, when the foundation of classical physics, Newton's Theory of Gravitation and Mechanics, was being challenged by a new theory, Einstein's General Theory of Relativity. This new theory stated that deductions from Newtonian mechanics only fitted the facts approximately, and that predictions from both theories were indistinguishable for situations we have so far examined. However, for cosmic phenomena where great masses and velocities are involved this difference in predictions may be detected by our 'crude' instruments.

The most important eclipse of the sun in the history of science occurred on 29 May 1919. It was a moment of magic, when our accepted perception of the whole universe was overturned  This eclipse provided the conditions for the crucial experiment. It was very delicate and involved photographing and measuring the amount of bending of a ray of light from a star when it just grazed the edge of the sun. Einstein's theory predicted a bending of 1.75" of arc while Newton's theory predicted 0.87" of arc.

To decide between the two theories, the British sent expeditions to two locations, in W. Africa and Brazil, during the 1919 eclipse of the sun. From their photographs they reported a displacement of 1.98" and 1.61". Later, during the 1922 eclipse an American team found displacements of 1.72" and 1.82" using two different cameras.

From these four data points evaluate the two theories by statistical methods.

35. Do you recall saying that a glass of beer would relax me, calm my nerves, steady my hand, sharpen my eye, improve my aim and allow me to shoot more baskets? I didn't believe you so I tested your claim by direct experiment.

My long term average in basketball free throwing is 20 baskets out of 25 attempts. I then ran 16 tests of 25 throws each (400 throws total), drinking a glass of beer before each test, and this is the number of baskets made :

24, 25, 22, 16, 20, 20, 23, 22, 19, 22, 18, 21, 24, 19, 22, 19

My shooting average seems to have gone up, but before I recommend that the rest of the team drink beer just before a game I want to be sure that this drink is effective. In fact I want to be at least 95% sure that a drink of beer is effective.

What should I conclude?

36. You say that freshmen are better at basketball free throwing than me. NONSENSE! What's your overall average? 18 out of 25? All right, let me try, (off to the gym and back again two hours later) . . .

    Here are my scores: 15, 19, 18, 18, 17, 16, 15, 17, 18

    I see that this is a bit low, but I'm sure that it's accidental. What, you don't agree? How sure are you about your claim? Are you willing to bet $95 to my $5 that you are better.

37. At my regular gas station I get 12 mpg in my giant SUV. But recently I drove to a further-away gas station and tried three fill-ups with another brand of gasoline, which gave 14 mpg, 17 mpg, and again 17 mpg. I think that these 2 gasolines should be equally good. What could I reasonably conclude at the 5% significance level?

**In the following problems use the 5% significance level**

38. My theory predicts that when I advance the spark plug timing, $Y$, in my test car the gasoline mileage of the car, $x$, goes down, or that $Y = -x$. So let me try 8 different spark plug settings, and for each measure the increase or decrease in mileage (mpg).

| $x$ (spark plug settings) | -4 | -3 | -2 | -1 | +1 | +2 | +3 | +4 |
|---|---|---|---|---|---|---|---|---|
| $\Delta y$ ( actual mpg ) | -3 | -4 | -3 | 0 | +2 | +1 | +2 | +5 |

   a) Evaluate the Least Squares line (the LS line) for my mileage data y.

b) Does changing spark plug settings affect the mileage of my test car?

Note: Very simple numbers are used in these problem

39. Given the following 3 data points

| x | y |
|---|---|
| 0 | 3 |
| 1 | 0 |
| 2 | 1 |

a) Find the least-squares line through this data
b) Do we have any reason to suspect that the slope is different from zero?

40. Given the following 12 data points

| x | y |
|---|---|
| 0 | 3, 3, 3, 3 |
| 1 | 0, 0, 0, 0 |
| 2 | 1, 1, 1, 1 |

Do you now suspect that the slope is different from zero?

41. Given

| y | 3 | 1 | 9 | 7 |
|---|---|---|---|---|
| x | 0 | 2 | 6 | 8 |

(a) Find the least squares line
(b) Does a change in x affect y?

42. Given

| y | 2 | 2 | 0 | 0 |
|---|---|---|---|---|
| x | 0 | 1 | 2 | 3 |

(a) Find the least squares line
(b) Does a change in x affect y?

CHAPTER **8**

# DECISION THEORY

*"The essence of knowledge is, having it, to apply it"*
*—Confucius (551-479 BC)*

In life we have to make all sorts of decisions. For example, next month I will have to go from San Francisco to Los Angeles. Should I fly or drive? Here are some thoughts.

Flying is so convenient and quick. But occasionally there are delays, but very rarely is there a fatal accident. On the other hand, driving is tiring, more expensive, accidents are more frequent but are less serious and rarely fatal.

If I'm a pessimist I'll surely be terrified that the plane will crash. But if I'm an optimist, I'll brush aside those silly fears.

There are also other considerations. For example the probability of being on a plane which has a bomb-carrying passenger is about a million to one. Does that scare me? If so then I may not want to fly. Or else, I may consider taking a bomb on the plane myself because the probability of having two bombs on a plane would then be only one in a trillion, and this may be so low as to assuage my fears.

As other examples

- If I slashed the price of gasoline at my gas station I certainly will sell more gas and maybe would make a greater profit. Should I?

- If I don't squeal to the cops I'll probably get about 10 years in prison. If I do squeal and finger our gang I'd probably beat the rap, but then I wonder what my good friends will do to me - fit me with a concrete suit or something similar?

Yes, living requires making lots of decisions and the action I take depends on my temperament - whether I'm a pessimist, an optimist, or whatever. Let's see how decision theory can suggest sensible courses of action whenever I have estimates of probabilities (P) and desirabilities or costs (D) of the outcomes of my actions.

## THE DESIRABILITY

The desirability, D, is difficult to define and quantify, and the best we can do now-a-days is measure it in terms of money. The legal systems of our societies have adopted this measure. For example:
- It's worth $10 to me to arrive at the big game on time, so I'll take a taxi instead of a bus.
- You punched him on the nose. That will cost you $1000.

There are other measures, for example the barter system:
- You desire a wife? This young girl will cost you three camels.

And then there is the biblical measure "An eye for an eye"
- You punched him on the nose, so he has a right to punch you.
- You killed one of his children. Now he has a right to adopt, free of charge, one of your children.

We adopt the monetary measure for desirability or (D). So now let us illustrate with a simple example how decision theory may help us come to a reasonable course of action.

### Should I drive my car to work, or should I take a bus?

Here are my best estimates of the daily costs and probabilities associated with going downtown to work.

**By car/day**

| | |
|---|---|
| Cost (gasoline, parking fee, auto depreciation) | $D = -\$10$ |
| Value to me to arrive on time | $D = +\$7$ |
| Annoyance of arriving late | $D = -\$3$ |
| Accident (insurance and aggravation) | $D = -\$1000$ |
| Probabilities: Arrive on time | $P = 0.9$ |
| Arrive late (frustration) | $P = 0.099$ |
| Accident once every 4 years | $P = 0.001$ |

## Desirability

**By bus/day**

| | |
|---|---|
| Daily cost (bus tickets) | D = - $2 |
| Arrive on time and then walk from bus stop | D = - $5 |
| Missed a bus connection and arrive late | D = - $10 |
| Probabilities:  Bus arrives on time | P = 0.8 |
| Bus is late | P = 0.2 |

*Let us tabulate all this information for easy reference*

| Action | Drive my car | | | Take the bus | |
|---|---|---|---|---|---|
| Cost $/day | -$10 | | | -$2 | |
| | Arrive on time | Arrive late | Have an accident | Arrive on time | Arrive late |
| Probability | 0.90 | 0.099 | 0.001 | 0.8 | 0.2 |
| Value to me, $/day | +$7 | -$3, traffic jams | -$1000 once every 4 years | -$5 | -$10, have to walk home from bus stop |

Let us list the four most useful rules for action

### Rule 1. The Optimistic.

I should choose the action which could lead to the most favorable outcome.
The best outcome by car - arrive on time    D = -$10 + $7  = -$3
The best outcome by bus - arrive on time    D = -$2 - $5  = -$7

*Therefore I should <u>drive my car.</u>*


### Rule 2. The Pessimistic.

I dread the worst, so I will stay away from the actions which may lead to the least favorable outcomes..
The worst outcome by car - an accident:   D = - $10 - $1000 = - $1010
The worst outcome by bus -  late:    D = - $2 - $10 = - $12

*Therefore I should <u>take the bus</u>.*

### Rule 3. <u>Mathematical Expectation, E.</u>

Here we look for the most desirable outcome in the long run. This means choosing the action which gives the highest (positive) expected value, a mathematical measure defined by

$$E = \Sigma(P\,D)$$

In this example consider 1000 days of going to work (about 4 years). The desirabilities to me are

$$\begin{aligned}
&\text{By car (D):} &&\begin{aligned}900\text{ days at:} &&-\$10 + \$7 &= -\$3/\text{day}\\ 99\text{ days at:} &&-\$10 - \$3 &= -\$13/\text{day}\\ 1\text{ day at:} &&-\$10 - \$1000 &= -\$1010/\text{day}\end{aligned}\\[6pt]
&\text{By bus (D):} &&\begin{aligned}800\text{ days at} &&-\$2 - \$5 &= -\$7/\text{day}\\ 200\text{ days at} &&-\$2 - \$10 &= -\$12/\text{day}\end{aligned}
\end{aligned}$$

Then the mathematical expectation is calculated to be:

For travel by car:   $E = 0.9\,(-\$3) + 0.099\,(-\$13) + 0.001\,(-\$1010) = -\$5$

For travel by bus:   $E = 0.8\,(-\$7) + 0.2\,(-\$12) = -\$8$

In the long run using the car is the more desirable way to get to work. It costs me $5 per trip versus $8 per trip by bus.

*So go <u>by car</u>*

### Rule 4  Minimize the Maximum Risk.

This is the conservative decision rule if you only have rough estimates of the probabilities of the various outcomes. For example, if you know that an accident is likely to occur once every 100 to every 1000 days then E, from Rule 3, is

$$E_{car} = 0.9\,(-\$3) + (0.09 \text{ to } 0.009)\,(-\$13) + (0.01 \text{ to } 0.001)\,(\$1010)$$

The worst outcome is for more frequent accidents. thus

$$E_{car} = 0.9\,(-\$3) + 0.09\,(-\$13) + 0.01\,(-\$1010) = -\$13.97$$

The expected value by bus is unchanged, from Rule 3,

$$E_{bus} = -\$8$$

Go *by bus*

### COMMENT

Rules 1 and 2 only use D values. while Rules 3 and 4 use both D and P values. So if you have estimates for P values then Rules 3 and 4 are more reliable approaches to decision making.

In long term situations where you can afford an occasional loss then Rule 3 will give you the most reliable long term result. If you cannot afford an unfortunate outcome then choose the conservative Rule 4.

## Problems

1. The TV weather report said that there was a 25% chance of rain tomorrow.

   If I guess correctly either way (I have my umbrella and it rains, or if I don't (and it's sunny) then everything is fine (Desirability D=0). But what if I guess wrong? If I do take the umbrella and it turns out sunny I'd be inconvenienced and feel silly (D = -3), however if I don't take it then I'd get wet and uncomfortable, etc (D = -10).

   *What should I do?*

2. I could drive, take a bus, or fly from San Francisco to Los Angeles, and here are my best estimates for the costs and benefits, D, and the probabilities, P, of the various expected outcomes:

   By car    Cost at $ 0.3/mile = $ 120
                 Arrive on time without incident, P = 0.95, D = + $30
                 Minor fender bender, P = 0. 05, D = - $500

   By bus    Cost = $90
                 Arrive on time, P = 0.8, D = - $20
                 Late arrival, P = 0.2, D = - $30

   By plane Cost = $150
                 On time arrival, P = 0.6, D = $100
                 (I don't want to miss an important meeting)
                 Late arrival, P = 0. 3999, D = -$20
                 Terrible accident, P = 0.0001, D = -$1000 000

   *How should I make the trip?*
   a) I have lived a clean life and so I am confident that God will smile on me and keep me safe.
   b) I'm jinxed and bad luck follows me everywhere I go.
   c) I make this trip to Los Angeles weekly and I want to do it most efficiently.

3. Our company is presently making plans for next year's investment on new chemical plants and two courses of action have been suggested.

The Gamblers recommend putting all our money into a large plant to produce a fluorescent nylon which would glow in the dark. If no other company has the same idea (30% chance of this happening) we will make a killing of $120M profit. However, if our competition gets the same idea, and they are much bigger than we are, then there would be a price war, we'd be squeezed out of the market, and overall we'd lose $40M.

The Timid Group suggests developing three smaller processes. If no one else duplicates any one of rhese these (20% chance) then we'd make $70M. If one is duplicated (50 -50 chance) then we'd still make a profit of $30M. If two are duplicated by our nasty competitors (20% chance) then we'd lose $20M. But if all three are duplicated then we'd lose a lot, $60M!

*What course of action should our company take?*

4. **The King chooses his Queen.** King Bumbledorf planned to marry so he ordered his Prime Minister to scour his kingdom to find 100 of the most beautiful, eligible damsels. He must then choose one of these ladies. His procedure was as follows:

He would test them one at a time (in random order). First he would test damsel number 1 for a day and a night and then decide to accept or reject her. If she is rejected she is returned to her village with a gift and can not later be reconsidered. Then the King would test damsel number 2, and so on, until he is satisfied. When he has chosen his bride, then the whole selection process is over because King Bumbledorf now has his Queen.

***Using Bumbledorf's procedure, how would you choose?***
*For help see "The Secretary Problem" on the internet.*

5. Extension to Problem #4

As reported on the internet, many American mathematicians, including Martin Gardner[*], say that King Bumbledorf insisted that he would ONLY marry <u>the very best</u> of the one hundred available damsels sent to his castle and that NO ONE ELSE would do - period. In addition, his decision to accept or reject a damsel was only based on the applicants examined up to that point. Remember, if he rejected a damsel she could not later be reconsidered.

I think that the results given by the mathematicians reporting on the internet are wrong...

[*] Phillip Yam, *A Tribute to Martin Gardner, 1914-2010*, May 22, 2010, Scientific American.

[*] Wikipedia, *http://en.wikipedia.org/wiki/Martin_Gardner*

CHAPTER **9**

# THEORY OF GAMES

Consider statistical decision theory. Here a person knows the probability and the desirability (the payoff) associated with the various courses of actions, and he then decides how to act according to some criterion.

But suppose he doesn't know what the probabilities are and suppose he has to decide how to act in a situation where someone else has an opposing interest, for example: poker, bridge, a love affair, car buyer and car salesman, Israeli and Palestinian.

The elements of these conflict are:

- Two or more interests at cross-purposes. This leads to 2-person, 3-person, etc. games

- Each person has some influence on the outcome of the conflict

The theory of games gives a model of such situations, and suggests how to act. How good are the predictions? This depends on how well the model fits the facts.


**1. Person or players.**

This is the number of conflicting interests.

    a) One person games - solitaire, should I drive my car to work or take a bus. These games are easy to solve, just calculate the mathematical expectation, and then choose the action to give the highest E value (use decision theory).

    b) Two person games.

    c) More than 2-person games - these are very hard and often impossible to treat.

We will focus upon 2-person games, and then touch on various extensions.

## 2. Strategies, payoffs, and the game matrix.

Consider two players, White (**W**) and Black (**B**). If **W** has three strategies and **B** has four, we have what is called a 3 x 4 game.

The **payoff** of a game is the amount that **W** (the person on the left - see below) wins when they play. Thus if **B** chooses strategy B2 and **W** chooses strategy W3 then the payoff is 7. This means that **W** gains 7 something, 7 candy bars, $7, or seven of whatever it is that they are playing for. Below we show the game matrix, the 3 x 4 strategies, and twelve payoffs.

|  | **Black** | | | |
|---|---|---|---|---|
| | B1 | B2 | B3 | B4 |
| W1 | 1 | 4 | 3 | 0 |
| W2 | 5 | 9 | 4 | 6 |
| W3 | 8 | 7 | 1 | 2 |

**W** wants to keep the payoff as high as possible and **B** wants it as low as possible

## 3. Zero-sum and non zero-sum games.

If $W_{payoff} + B_{payoff} = 0$ we have what is called a <u>zero-sum game</u>. This means that if one person loses $102, the other wins $102. If they do not win and lose the same amount then we have a much more complicated situation, <u>a non zero-sum game.</u>

## 4. Criterion.
How should you play? wildly? conservatively? Game theory suggests

> *Play so that you can gain as much as possible safely, even when playing against a skillful opponent.*

Of course, if your opponent plays stupidly you may be able to gain more.

If you do not want to use this criterion then go to another chapter. Game theory is not for you.

## 5. Search for a saddle point.

First set up the game matrix, then see if a saddle point exists. If it does give a cheer because that simplifies matters. Now let us see how to find the saddle point for the above 3 x 4 game.

- Start by considering **W**. Find the minimum payoff, the worst, for each of **W**'s three rows and write the three numbers on the right of the matrix.

- Circle the largest of these three numbers. This is the maximum of the minima and is called the <u>maximin</u>.

- Now for **B** find the maximum payoff for each of **B**'s four columns and write them below the four rows.

- Next circle the smallest of these four numbers. This is called the <u>minimax</u>.

We show this procedure in the sketch below.

|       | Black |   |   |   |    |                      |
|-------|-------|---|---|---|----|----------------------|
|       | 1     | 4 | 3 | 0 | 0  | ← minima of the rows |
| White | 5     | 9 | 4 | 6 | (4)| ← maximin            |
|       | 7     | 5 | 1 | 2 | 1  |                      |
|       | 7     | 9 |(4)| 6 |    | ← maxima of the columns |
|       |       |   | ↑ |   |    |                      |
|       |       | minimax |   |   |    |                  |

If the **minimax = maximin** we have a saddle point. In that case **W** plays that particular row (W2), and **B** plays its particular column (B3), and the value of the game, the payoff, is 4. Thus **W** gains 4 every time **B** and **W** play. In this situation:

- Both players know the payoff matrix
- If either player deviates from that strategy his payoff will decrease.
- Each player can inform his opponent how he will play his next move. It won't hurt him. His strategy is "transparent".
- This is called a <u>pure strategy</u>.

The preliminary search for a saddle point should be done for all zero-sum games - whether 2 x 2, 2 x m, or m x n, or 3 x 4, as in the example on the previous page.

When no saddle point is found then we have to use a mixed strategy, sometimes playing W1, sometimes W2. In this situation we will only consider 2 x 2 games. Other games become too complicated.

**6. 2 x 2 zero-sum games which have no saddle point.**

Consider the following game matrix which has no saddle point.

**Black**

|  | B1 | B2 |
|---|---|---|
| W1 | a | b |
| W2 | c | d |

White

In solving the tricky mathematics we find that **W**'s optimal strategy is to play W1 and W2 in the ratio :

$$\frac{W1}{W2} = \frac{|c-d|}{|a-b|} \quad (1)$$

Similarly **B** should play in the ratio

$$\frac{B1}{B2} = \frac{|b-d|}{|a-c|} \quad (2)$$

So the % of time that **W** should play W1 is

$$\% W1 = \frac{100 |c-d|}{|a-b| + |c-d|} \quad (3)$$

and the % of time to play W2 is

$$\% W2 = \frac{100 |a-b|}{|a-b| + |c-d|} \quad (4)$$

Similarly the percentage of time that **B** should play B1 is

$$\% B1 = \frac{100 |b-d|}{|a-c| + |b-d|} \quad (5)$$

and for B2

$$\% B2 = \frac{100 |a-c|}{|a-c| + |b-d|} \qquad (6)$$

The value of the game or how much **W** wins or the payoff of the game is:

$$W_{payoff} = \frac{|c-d|\,a + |a-b|\,c}{|c-d| + |a-b|} \qquad (7)$$

As for **B**, if **W** wins \$X then **B** loses \$X. So...

$$B_{payoff} = - W_{payoff} \qquad (8)$$

Remember that this is a zero-sum game.

## Example 1  Rumanian coin flip

Hey, hey, you freshman, how about coming up to my place for an evening of sarsaparilla and Rumanian coin flip. What!? You've never heard of that game? Well it goes like this: We, both at the same time, shout either "heads" or "tails".

- If we both say "heads" (H), I will pay you \$3
- If we both say "tails" (T), I will pay you \$1
- If we don't say the same thing (HT or TH), the only fair thing is that you pay me \$2

You're coming? Good, and by the way be sure to bring lots of money with you.

*How should I (**W**) play and what would be my payoff?*

*How should you (**B**) play, and what would be your payoff?*

## Solution

Let W1 and B1 stand for heads
and W2 and B2 stand for tails
Then the game matrix is .........

|    | B1 | B2 |    |
|----|----|----|-----|
| W1 | -3 | +2 | -3 |
| W2 | +2 | -1 | -1 |
|    | +2 | +2 |    |

(me on left side; you on top)

In checking for a saddle point we find that the minimax is not equal to the maximin, because $-1 \neq +2$. This shows that there is no saddle point, and that we will have to use mixed strategies to get the most out of this competition. From Eqs. (1), (2), (7) and (8) we find:

$$\frac{W1}{W2} = \frac{|c-d|}{|a-b|} = \frac{|+2+1|}{|-3-2|} = \frac{3}{5}$$

$$\frac{B1}{B2} = \frac{|b-d|}{|a-c|} = \frac{|+2+1|}{|-3-2|} = \frac{3}{5}$$

$$W_{payoff} = \frac{|+2+1|(-3) + |-3-2|(+2)}{|+3+5|} = \frac{3(-3) + 5(+2)}{8} = \frac{+1}{8} = 12.5 \text{ cents}$$

and since this is a zero-sum game

$$B_{payoff} = \frac{-1}{8} = -12.5 \text{ cents} \quad \text{or} \quad B \text{ loses } 12.5 \text{ cents/game}$$

So we both should randomly play three heads and five tails on the average. With this strategy Eq. 3 shows that on the average I will win $0.125/game. **Note:** the freshman may know the ratio of heads and tails that I plan to use, but he must not know what my next move will be.

However he should be warned that if he deviates from that 3:5 strategy then he may lose more if I'm clever, which I am.

### 7. 2 x 2 Non zero-sum Games - Grand Opera

These are fascinating games which often represent simple pictures of real conflicts between people, such as Shiite vs Sunni, Israeli vs Palestinian, Serb vs Albanian. Although these games are idealizations,

# Non-Zero-Sum Games

they may give valuable insights into possible resolution of these conflicts.

In these games there is no systematic route to a payoff, but game theory does tell you what is a sensible course of action and what is not.

These games are best described in terms of an example. Let us consider the well known Puccini opera "Tosca".

> The conflict is between *Tosca*, a beauty, and *Scarpia* the nasty police chief. *Scarpia* lusts for *Tosca* and comes up with a clever plan. He first imprisons her lover, Cavaradossi, and then has it arranged that he be executed, by firing squad the next day at dawn.
>
> *Scarpia* then slyly offers *Tosca* a deal: he would let Cavaradossi live if she would sleep with him this night and let him ravish her. *Tosca* responds saying that she would agree but only after he issues an irrevocable order to the firing squad to fire blanks and then release Cavaradossi.

Now what are her choices and what are his choices?

- Should she be honest and let him have his way with her, as she had agreed upon (**H** = honest), or should she double cross him, stab him in the neck with her hairpin, paralyze him and gloat to see him slowly die (**D** = dishonest).

- He also has a choice, either to honor his promise and spare the life of and release Cavaradossi (**H**), or double cross her and get rid of the competition (**D**).

Let us set up the game matrix for her and for him, and let the numbers -2, -1, +1, and +2 represent progressively better outcomes. Thus -2 is the worst, and +2 is the best outcome, and let $x$ refer to *Tosca's* payoff, and let $y$ refer to *Scarpia's* payoff. Then their combined game matrix is:

|  |  | *Scarpia's* choice ($y$) | |
|---|---|---|---|
|  |  | **H** | **D** |
| *Tosca's* choice ($x$) | **H** | $x_1, y_1$ | $x_2, y_2$ |
|  | **D** | $x_3, y_3$ | $x_4, y_4$ |

*How should we fill in **Tosca's** and **Scarpia's** payoffs?* Thinking about it we get:

- If **Tosca** and **Scarpia's** are both honest $(x_1, y_1)$
  she has one bad night but got her lover back ($x_1$ = +1), while he enjoys her for the night but loses her from then on ($y_1$ = +1).
- If **Tosca** is dishonest and **Scarpia's** is honest $(x_3, y_3)$
  she has killed the beast and got her lover back ($x_3$ = +2), but he has lost everything, even one good night and dies ($y_3$ = -2)
- If **Tosca** is honest and **Scarpia's** is not $(x_2, y_2)$
  she loses her purity and her lover ($x_2$ = -2), while he has had a good time and has gotten rid of his competitor forever ($y_2$ = +2).
- If **Tosca** and **Scarpia's** are both dishonest $(x_4, y_4)$
  she kills the beast but saved her purity ($x_4$ = -1), and he dies while knowing that he has killed her one and only ($y_4$ = -1)

The overall payoff matrix (8 items) is then

Payoff for **Scarpia** ($y$)

|  |  | H | D |
|---|---|---|---|
| Payoff for **Tosca** ($x$) | H | +1  +1 | -2  +2 |
|  | D | +2  -2 | -1  -1 |

Now if **Tosca** and **Scarpia** only want their very best outcomes, +2, +2, they both will double cross the other. The result will be -1 and -1 and they both will lose.

If one is honest, the other dishonest the honest party loses, -2, while the dishonest one gains, +2.

However if they both trust each other then both gain somewhat, +1 and +1.

What happens in the opera? Tragedy, of course.

**Conclusion.** This example shows the value of compromise. In trust, both parties gain somewhat, +1, though not everything they want, +2, while decisions based on self interest alone can lead to disaster -2.

This could well apply to the many problems between peoples of today, especially political problems. Think of the Israelis and the Palestinians. Each has wanted +2 for over 50 years.

**8. Games with more than 2 Persons.**

Suppose that 3 people **A**, **B**, **C**, have to divide $1200 by majority vote, with bargaining allowed. At first they decide to be fair and to divide the money evenly, $400 to each of them. Then **A**, thinking only of his own interest, approaches **B** with a tempting suggestion, to divide all the money between them, and cut out **C** completely. That way each receives $600.

Not a bad idea thinks **B**, but before they take a vote, **C** not being a dumb cluck, approaches **A** saying "I've got a better deal for you, $800 for you and $400 for me, and a big fat zero for that crook **B**."

But then **B** approaches **C** saying "Why should that greedy **A** get $800? Look, I've got a proposal for you, ... ". And so it goes, on and on.

In fact there is no solution to this problem, and this shows the value of compromise, for acting for the joint good of all three people, and not just trying to maximize the good for yourself.

*In general, for n-person games, if all parties act cooperatively, aiming for a compromise they will all do fairly well, and you get a stable solution, or an "equilibrium" as it is called in game theory.*

However, if any of the parties wants it all, or if coalitions form (as in the dividing of the $1200) then there is no equilibrium.

**Von Neumann** and **Morganstern**, in their pioneering book [3] which introduced the whole concept of game theory worked out the solution to two-person zero-sum games.

Nash extended these concepts to deal with n-person games where the parties cooperate. However, when each person tries to maximize his profit, Nash showed that no equilibrium exists.

If you want a good shot at a Nobel prize tell us how to treat those games where the parties are only concerned with their own self interest.

## 9. The Finite and the Infinite Universe.

Let us introduce the well-known problem called "the Tragedy of the Commons" (see G. Hardin, *Science*, 1243 -1248, 13 Dec. 1968)

> A village has a common field that anyone can use, for picnics, for playing games, whatever. This field does not belong to anyone, it is just part of the village.
>
> *If you take advantage of this 'free' grass and graze your cow on the commons it will get fat and healthy, you will become a bit more prosperous, and what will be the harm?* Negligible. But if everyone does this the field will become overgrazed and trampled and will end up as a muddy mess. So everyone loses.

**Here are other examples.**

- Suppose I have to put in a septic tank and drain field when I build my house, and that will be very costly. But the river runs by my house. *What is the harm if I just run my drain pipe into the river and forgot all that nonsense about septic tanks?* Negligible. Yes, but what if all the people in town did it?
- Historically when white man tramped across North America he viewed it as infinite. Shoot as many bison as you wish, for pleasure. Shoot deer and game animals. Don't worry because the number available is limitless.
- Pump the petroleum out of the ground and use it carelessly. We've got lots of it. Today it is even cheaper than bottled water! But we are beginning to see different attitudes.
- The latest method of fishing has two giant trawlers crossing the ocean a mile apart and dragging a three mile curtain-like net between them. This net sweeps up all the fish, large and small, small whales, porpoises, turtles and everything else. It is very profitable.

The answer to these questions hinges on whether we consider the environment to be finite or not - whether the field is so big, the river so big, the country so big that what we do does not affect it significantly.

We may argue and disagree about fields and rivers but in our minds the oceans are still viewed as infinite. Just discard whatever you want into it. European nations have been using the seas off their coasts to rid themselves of tons and tons of hot radioactive wastes. What a cheap solution to their waste disposal problem. But there are a few spoilsports who say "Wait, why do you think it reasonable to assume that the ocean is infinite in size?" Let us check this assumption.

**Let us make some calculations.**

Suppose you dump a 42 gallon barrel of radioactive waste into the ocean anywhere, mix this marked waste with all the waters of all the oceans of the world (the Pacific is on average more than 4000 m deep) and then scoop a teaspoonful of ocean water from anywhere in the world and analyze it. *What do you think is the probability of finding a single molecule of that original barrel of radioactive waste?*

Well, you may be surprised to learn that that teaspoonful of water would be teeming with about 25 000 of those toxic molecules! The same with every other teaspoonful of ocean water everywhere in the world - *incredible*.

**So we should not consider that the oceans of our world are infinite.**

So, whether an action (letting a cow graze, to dump waste in a river, etc.) is harmful or not, can be done without feeling guilty or not, depends on whether the "universe" we are dealing with can be considered to be finite or not.

**10. The value of Game Theory** is that it lays bare the different kinds of thinking that apply or should apply to different kinds of conflicts. It makes us face the questions of values and consequences. Here is a final thought: the economics guru **Adam Smith** [in The Wealth of Nations (1776)] wrote:

> *that an individual who "intends only his own gain" is "led as if by an invisible hand to promote the public interest"*

However, people may forget that he also said:

> *that this requires perfect competition of many separate interests, no one large enough to influence the overall outcome (hence no monopolies)*

This is the philosophy that guides today's free market societies. In a nutshell he says:

> *An individual should look out for his own best interest.*
> *This way the interest of all will be maximized.*

I wonder how this view squares with that of game theory? What do you think?

**Here's a fable to end this discussion.**

It concerns a little community of 35 families living in the village of Gleneden, nestled in the heart of Emerald Valley. The community was so impoverished that last New Year's eve party was a sad affair with just a few drops of whiskey per person.

Right after the party Paddy had a brilliant suggestion. Let each family each Sunday donate a thimbleful of whiskey and pour it into the community cask, which was protected by the parish priest. By next New Year's day the cask would be full and what a grand party the village could then have.

The village enthusiastically embraced this idea and throughout the year the villagers followed the filling of the cask with increasing excitement.

Sure enough by New Year's eve the cask was at the point of overflowing, The party began, drinks were poured, glasses were filled, the partiers gave a big cheer, shouted 'bottoms up' and then they downed their glasses of pure *'water'*!

## 11. All sorts of Conflicts and Game Theory.

John von Neumann [4] opened the door to game theory by developing the basic theorem for treating two-person zero-sum games. Its application was the basis of operations research, a really important tool which helped win the Second World War.

John Nash generalized von Neumann's theorem to several (not just two) players acting cooperatively, and he received the 1994 Nobel prize for it.

Game theory deals with all sorts of conflicts beyond these ideal situations, and it has found important uses in - economics, politics, religion, gambling, advertising, biology, etc.

*Let me illustrate this with a few diverse examples.*

**12. The Shubik Dollar Auction [2].**

Here is a game you can try with a group of friends. Often you will get surprising results.

> A dollar bill is put up for auction. The minimum bid is 1 cent and continues by the usual rules of auction with one exception - the auctioneer is paid by the highest bidder, but also by the second highest bidder. The highest bidder gets the dollar bill, the second highest bidder gets nothing. In this game the bidding usually starts casually, but as the bids rise the bidders find themselves hooked.

Many real life situations have the same characteristics as this game, for example:

> President Johnson's dilemma. In 1964 he expanded the war in Vietnam just a little bit, and then a little more (for democracy, freedom, justice), but by 1968 he just couldn't withdraw (national honor, avoidance of weakness).
>
> Taxi or Bus. The longer you wait at the bus stop for the bus to come, the harder it is for you to decide to take a taxi.
>
> TV Commercials come more frequently close to the end of a TV show, when you've already invested time watching the show.
>
> People in unhappy jobs. Employees are reluctant to quit their jobs because of accumulated retirements benefits, stock options, etc.

**13. The Prisoner's Dilemma.** This is a curious problem, probably the most famous in game theory. It has been intensively studied, wrestled with and written about. It goes like this:

> Two scruffy fellows were stopped while speeding away at 180 km/hr from the scene of a serious crime. The police have no solid evidence to implicate them in the crime, but are pretty sure that they did it.

Unfortunately the police can only prove excessive speeding. How to get a confession? Here is the police strategy. First, the two suspects are lodged in separate cells. Then the prosecutor approaches one of them and offer a plea bargain suggestion:

"If you confess and implicate you buddy I will set you free, and I will forget your speeding ticket; however your buddy will get 10 years. By the way I will make the same offer to your buddy. Of course if he also confesses then your confession is not worth much as we know everything without it. In that case you'll get 5 years each. If neither of you confesses, all that I can do is get you on the 180 km/hr speeding ticket which will cost you a year in jail."

"Think of this. I'll come back tomorrow to learn of your decision."

*Write a game matrix for this situation. Then tell what you would do if you were one of the two prisoners.*

### 14. Chicken.

This is a simple game that lays bare the key to many conflicts. The name "chicken" comes from the 1955 Hollywood movie "Rebel without a Cause".

Two teenagers drive stolen cars towards each other on a narrow road. The first to swerve out of the way is "chicken" and is shamed by the gang.

*Numbers 15 and 16 are examples of this type of conflict.*

**15. The Cuban Missile Crisis** in 1962 between the US and the USSR is an example of a chicken game. President Kennedy persuaded Chairman Khrushchev that the US would not shrink, even from nuclear war. In the end Khrushchev backed away.

*Who do you think was the rational leader?*

### 16. World War Two.

Before the Second World War England's prime minister Neville Chamberlain was unwilling to risk the worst - war; and Hitler won quite a few chicken games against him - Sudetenland, Czechoslovakia and Austria. In the end Chamberlain recognized the situation and forced Great Britain to go to war.

**Note**: It looks like Khrushchev versus Kennedy, and Chamberlain, early on, versus Hitler, did not make the same decision. *Discuss this.*

Also Israeli versus Palestinian seem to be embroiled in such a game today.

### 17. The *Science - 84* Game.

To gain circulation *Science - 84* planned a game in which their readers could apply for $100 or for $20, all this for the cost of a posted letter. If fewer than 20% of the applicants applied for the $100 then everyone would receive what they applied for. However, if more than 20% applied for the $100 then no one would receive anything. The editors eventually got cold feet and decided not to publish the game.

*If the game was published would you play it, and if so, how?*

### 18. The Smallest Number Game.

Another scheme to gain circulation of a magazine goes as follows. Choose a positive integer, write it on a postcard and send it to the magazine. The winner is the player who sends in the smallest number which was not sent in by anyone else.

### 19. The Least Popular Number Game.

Here's another variation. Choose an integer between 1 and 20, write it on a post card and send the card to the journal. The persons who sent in the least popular number wins.

### 20. The Million Dollar Game.

The US Postal Service profits from First Class mail and wants to encourage people to use this service. Here is one way:

Offer a million dollar prize, and advertise it in all the post offices around the country. All you do is just send in to the US Post Office an envelope with a sheet of paper with your name and address. If you are the only applicant, you win the whole ONE MILLION dollars, tax free - it's that simple. If two apply then the winner, chosen by lot, gets half a million. If ten people apply then the winner receives just $100 000, and so on.

*If you were the postal commissioner what would you think of this scheme?*

## SUMMARY

It seems that there are a number of factors to consider if we are to analyze conflicts. Is it a:

- zero-sum game or not?
- two-person or n > 2 person game?
- finite or infinite system?
- pure hate or good faith conflict?

## REFERENCES AND INTERESTING READINGS

1. L. Mero, *Moral Calculations*, Copernicus, Springer-Verlag, 1998.
2. M. Shubik, "The dollar auction game: A paradox in non cooperative behavior and escalation." *Journal of Conflict Resolution*, 15, 109 (1971).
3. J. von Neumann and O. Morgenstern, *Theory of Games and Economic Behavior*, Princeton University Press, Princeton, NJ, 1947.
4. P. M. Morse and G.E. Kimball, "How to Hunt a Submarine", *The World of Mathematics*. J. R. Newman, Ed. Simon and Schuster, New York, NY, 2148, 1956.
5. J. Maynard Smith, and G.R. Price, "The Logic of Animal Conflict", *Nature*, 246, 15 (1973).
6. J. D. Williams, *The Compleat Stategyst*, McGraw-Hill, 1966.
7. S. J. Brams, *Biblical Games*, MIT Press, Cambridge MA, 1980.

## PROBLEMS

### 1. Bulgaria's national game.

Bulgaria has its own version of the Rumanian coin flip called, naturally, the Bulgarian coin flip, which they love to teach tourists. In this game:

- If we both say heads (H), I pay the Bulgarian $7.00
- If we both say tails (T), I pay him $1.00
- If I say H and the Bulgarian says T then he pays me $3.00
- If I say T and the Bulgarian says H then he pays me $5.00

*I don't think I should play this game. What do you think? But if I do what should be my strategy, and on the average how much would I win or lose on each play?*

### 2. Bombers and Fighters.

Two bombers are needed for this mission, one to deliver the bomb, the other for support. The first bomber's tail is well protected by the second bomber, while the second one's tail is more vulnerable.

From past experience we know that the enemy has a 20% chance of bringing down the first bomber, but a 60% chance of bringing down the second.

*a) Which bomber would they likely attack, the first or the second, and*
*b) Which should be carrying the bomb?*
*c) What is the probability that the bombing attack will be successful?*

### 3. The Wild West.

The way he studied me in the saloon this afternoon as he drooled and filtered tobacco juice through his beard into his beer mug convinced me that Flapjack Fandango would try to intercept the payroll that we are to deliver. So we'll split up and take different routes and this way he won't know who is carrying it.

If he tries to follow me out of town and then intercept me on my speedy steed my chance of getting through is 4 in 5; but if he follows you on your old nag you will have 1 chance in 5 in getting through.

Now, what is important is that the payroll go through, and a few bullets here and there don't really matter. I also am quite sure, by the way he plays poker, that he's a sharp cookie and can figure the odds as well as anyone else.

  a) *I wonder, who should carry the payroll?*
  b) *Who will he follow?*
  c) *What is the probability that the payroll will be delivered successfully?*

### 4. Pussy's Lunch.

A hungry cat sits atop a field mouse burrow which has two exits, one on the East side and one on the West side. If he watches the East exit and a mouse tries to escape on that side the cat will have an 80% chance for lunch. If the mouse leaves on the other side his chance of escaping is 80%. If the cat watches the West exit his chance of catching an escaping mouse is only 60% on that side, but 10% if the mouse tries to escape on the other side.

### 5. Pointing Fingers.

While resting in a Las Vegas lounge a friendly fellow came up to me and asked if I'd play a little game with him. It goes like this

- He and I point one or two fingers at each other.
    - If we both point one then I pay him $5.
    - If we both point two I pay him $30.
    - If he points two and I point one he pays me $15.
    - If he points one and I point two he pays me $20.

    *How should I play and how much can I expect to win in the long run?*

6. **Stone - paper - scissors** is a well known child's game.

    Write its game matrix, and without any complicated mathematics, tell how you should play against a shrewd opponent, and tell what should be your payoff.

7. **Operations Research.**

    OR had its beginning during World War Two. Here is an example where it saved many, many lives and maybe the war itself.

    Convoys of 100 or more ships regularly crossed the Atlantic from the US to serve as the lifeblood for the British Isles. German submarines in "wolf packs" tried to intercept and destroy these ships and thereby bring Britain to its knees.

    Now these convoys had a choice of two broad routes; first, the more or less direct southern routes; and second, the northern routes which skirted Greenland and then wandered through the miserable foggy iceberg laden waters of the Northern Atlantic.

    Ships on the southern routes could be easily intercepted and attacked, but these routes were frequently patrolled by British anti-submarine aircraft. Some ships would be lost, but the subs may not get away free (overall -8 for the convoy). Ships on the northern routes were hard to find, but had no protection from the air, and when found it was disaster for them (-12 for the convoy). If the convoy took one route, while the "wolf pack" lay in wait in the other, then there would be no contact (0 for the convoy).

The subs had to choose whether to lie in wait along the southern routes, or to wander and even get lost in the foggy northern waters.

> *What route should the convoys take, and where should the subs lie in wait?*

**8. Repeat problem 7 with one change.**

Even if the convoy took the northern route it usually spelled trouble even without meeting submarines - collisions with icebergs and with other ships in dense fog, and even occasional lost ships (-4).

**9. Decision and Game Theory.**

In studying for my final I have two choices:

- concentrate on statistics (S)
- make a general review (G)

I have time only for one of these, not both.

**Choice 1**   If I concentrate on statistics and this is what the final is about then I have an 80% chance of passing the course, but if I guess wrong then my chances drop to 40%.

**Choice 2**   If I concentrate on a general review and if the final emphasizes this then I have a 60% chance of passing, but if the final is on statistics then I am sunk with only a 20% chance of passing.

> (a) *From last year's freshmen I get the impression that the odds are 3:1 on statistics being asked. If it is true what should I study and what are my chances of passing the course?*

(b) *If the prof is mean and knows the odds and is out to get me what is he liable to ask, what should I study, and what are my chances of passing the course?*

10. **Zero-sum/Non zero-sum.**

Under what conditions would you consider the following conflicts to be reasonably represented by zero-sum and by non zero-sum games. Discuss.

- Husband and wife play chess.
- Two male lions fight for their pride.
- An army patrol skirmishes with the enemy.
- An antelope meets a crocodile.
- The dispute between Palestinians and Israelis.

11. **Compare Decision and Game Theory in Solving a Problem.**

Let's play a game which goes like this.

You and I shout either "heads" or "tails" (H or T) at the same time

- if we both say H, I pay you $ 5.00
- If we both say T, I pay you $1.00
- If we don't say the same thing (HT or TH) you pay me $3.00

a) *I plan to crush you using decision theory by shouting H or T randomly and equally frequently. How much would I win on the average no matter what you do?*

b) *I wonder if there is a better strategy? Would game theory help, and if so, how much can I expect to win using it?*

**12. Here's a puzzling situation**

"if something belongs to everyone,
then it doesn't belong to anyone"

*How does this idea apply to a nation's National Parks?*

# Chapter 10

# Galileo and Newton
# Kinematics and Dynamics

*Educators are in the business of forming minds, not filling them.*

Since ancient times man has tried to figure out how things around him worked. This knowledge could add to his power of action and could help him defend himself against possible dangers. In the Western World the first recorded book to consider such matters was written by Aristotle about 500 BC. His ideas became the dominant concepts for the teachings of European scholars and churchmen for about 2000 years. This was a remarkable book, to have influenced the thinking and teachings for such a long period of time. But it included all sorts of nonsense. Let us illustrate the sort of thinking of the medieval scholars who followed Aristotle's teachings with a couple of examples.

**Moving Bodies**  Consider what happens when I throw a stone through the air. One explanation says that when the stone leaves my hand it is not pushed by my hand any longer so it should stop dead! How come it still goes on traveling? Can it remember that I had pushed it? No, it cannot, so this is puzzling.

Well, Aristotle's guiding principles of motion are that

*"nature abhors a vacuum"*

and that

*"a continually acting force is needed to maintain motion"*

and these two laws are the key to a reasonable solution to this problem, don't you see? Because as the stone leaves my hand it pushes air out of the way, a vacuum is created behind the stone, air rushes in and gives the stone a push. So it is pushed on and on. That's how it moves.

You may criticize this argument by saying that if you threw a stone in a vacuum no air rushes in behind the object to push it forward, so no thrown object can move. The Aristotelian scholar replies haughtily that this is a nonsensical objection because another of Aristotle's laws says that nature does not allow a vacuum to exist ( see above).

Another explanation, this by Leonardo da Vinci, says that the stone gets something which he calls a *forza* which causes it to move. He explains it better than I can so let me quote him:

> "I say that the *forza* is a spiritual quality, an invisible power which, by means of an external and accidental violence, is caused by the motion and introduced, fused, into the body; so that this is enticed and forced away from natural behavior.
>
> The *forza* gives the body an active life of magical power, it constrains all created things to change shape and position, hurtles to its desired death and changes itself according to circumstances. Slowness makes it powerful and speed, weak - it is born of violence and dies in freedom. The stronger it is the more quickly it consumes itself. It furiously drives away anything that opposes it until it is itself destroyed - it seeks to defeat and kills anything that opposes it and, once victorious, dies.
>
> It becomes more powerful when it meets great obstacles. Everything willingly avoids its death. All things which are constrained constrain themselves. Nothing moves without it. A body in which it is born does not increase in weight or size. No motion that it creates is lasting. It grows in exertion and vanishes in rest. A body on which it is impressed is no longer free."
>
> — Leonardo da Vinci, from R. Degas, *The History of Mechanics*, translated from Latin and Italian by J. R. Maddox, pg. 78, Dover, 1988.

Leonardo spent quite a few words telling you what he meant by *forza*. What do you think he meant? In today's terms what was he talking about — force, moment, momentum, inertia, kinetic energy, or what?

These explanations are examples of typical rationalist reasoning (see Chapter 1). Scholars tried to reason out the law of projectiles (a very important problem in those days when armies lobbed rocks and containers filled with hot oil at each other). They did no experiments to check their statements.

**Falling Bodies**  Let us look at one more problem, that of falling bodies.

Aristotle said that heavy bodies fall faster than light ones - twice as heavy, twice as fast - ten times as heavy, ten times as fast. Let us consider this idea.

Suppose that at a given instant we drop three weights, a 2 lb, a 4 lb, and a 6 lb weight from the top of the leaning tower of Pisa. According to Aristotle the instant when the 6 lb hits the ground the 4 lb is 2/3 the way down and the 2 lb is 1/3 the way down the tower.

Now let me ask you, if I tied a 2 lb weight to a 4 lb weight and threw this pair off with the other weights where would it be? (see figure below)

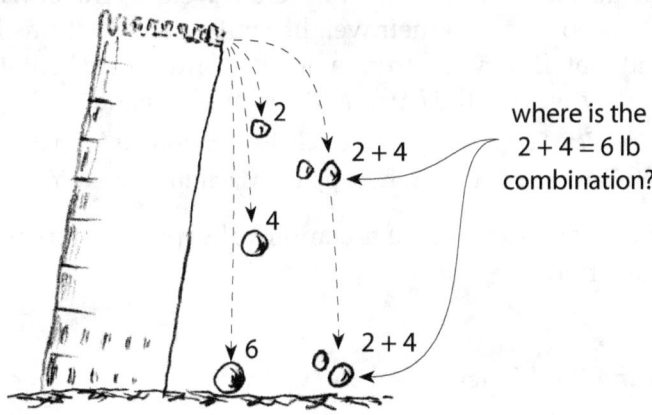

Now, Galileo questioned this 2000 year old truth somewhat differently - He did not rely on reasoning. He did experiments, and from these he concluded that all these weights fell at nearly the same rate. However, he had to admit that the 6 lb weight did, in fact, reach the ground first as stated by Aristotle.

I could see the Aristotlers chuckling and saying: "Aha, you are wrong because the heavy weight hit the ground first, as predicted by Aristotle". What was Galileo's answer to Simplicio, the character in his story, who was defending the Aristotelian position? From Galileo's *Two New Sciences*, translated from French by H. Crew and A. de Salvio, pg. 64, Macmillan, New York, 1914, we read of Galileo saying:

> "But, Simplicio, I trust that you will not follow the example of many others who divert the discussion from its main intent and fasten on some statement of mine which lacks a hair's-breadth from the truth and, under this hair hide the fault or another which is as big as a ship's cable. Aristotle says that "an iron ball of 100 pounds falling from a height of 100 cubits reaches the ground before a one-pound ball has fallen a single cubit.
>
> I say that they arrive at the same time. You find on making the experiment, that the larger outstrips the smaller by two finger-breadths. Now you would not hide behind these two fingers the 99 cubits of Aristotle, nor would you mention my small error and at the same time pass over in silence his very large one?"
>
> Aristotle declares that bodies of different weights travel with speeds proportional to their weights; but I claim that this is false and that, if they fall from a height of fifty or 100 cubits, they will hit the earth at the same moment."
>
> —from the *"Two New Sciences"* by Galileo Galilei, translated from French by H. Carew and A. de Salvio, pg 64, Macmillan, New York, 1914.

Here we have a good example of an empirical explanation trumping a rational explanation.

## Kinematics and the Pioneer - Galileo Galilei

Modern science has in large part adopted a new method of attack upon the unknown which says that propositions arrived at purely by logical means say nothing about our world.

> *Science just does not accept the ideas and teachings of any accepted authority without experimental evidence.*

This method was hesitatingly attempted before the 16th century. Galileo however was the first to use it openly, consistently and effectively. As a result, to him more than any individual should go the credit of being the father of modern science.

His method was to ask, not WHY things happen as they do, but WHAT IS IT that happens. For example he felt that why moving objects kept moving was not the important question. Indeed, to this day we do not know why things set in motion keep on moving. We just know that they do. He studied how they move, how they fall and how they roll, and so on; and from this came the study of kinematics.

Consider a 2 lb, a 4 lb and a 6 lb weights falling from the tower. What seems like a reasonable approximation to start with? How about this one:

*All bodies fall at the same rate*

Now this is just my guess as to how Galileo thought. Anyway he arrived at the above guess. What else did he know? The following seemed true, that

*the further an object fell the greater its velocity.*
*Hence falling motion was accelerated.*

So now he had three choices to consider. The acceleration could be constant, increasing or decreasing. Galileo assumed the simplest case - then tried to check his assumption.

## GALILEO INTRODUCED KINEMATICS

First of all let us see what kinematics is and what Galileo did to merit being called the founder of modern science and physics. Physics, which studies the physical world, is a very broad subject and kinematics is just one of its parts. Kinematics makes use of two concepts which are about the most primitive of our concepts. They are:

1. **Concept of Space** The idea of space refers to our ability to perceive objects apart from each other. The study of relative positions of various (stationary) objects is the subject matter of geometry.

2. **Concept of Time** Our world is not motionless, objects change their positions and shapes with time. Thus in describing the world we have to consider describing events occurring one after the other. This is the subject matter of kinematics, hence kinematics is the study of objects in motion.

4. **Defined concepts** These are velocity and acceleration
   (a) Velocity - describes the motion of a point giving the direction of motion and telling how fast the motion is.
   (b) Acceleration - Any motion in which the velocity varies in magnitude, direction, or both (with time) is called accelerated motion.

**Derivation of Galileo's equations for uniformly accelerated motion**

Let us remember that Galileo guessed that the acceleration of a freely falling object was constant. Let us use this and let us call "a" the value of the constant acceleration, defined as

$$a = \frac{v_2 - v_1}{t} \quad \left[ = \frac{m}{s^2} \right] = \left[ \frac{\text{meters}}{\text{sec}^2} \right] \quad \text{(i)}$$

$$\text{or} \quad \boxed{v_2 = v_1 + at} \quad \left[ = \frac{m}{s} \right]$$

However Galileo and we today cannot measure velocities directly, only distance and time, so let us try to deduce another equation from (i) which we can work with.

Now
$$\bar{v} = \frac{v_1 + v_2}{2} = \frac{v_1 + (v_1 + at)}{2}$$

But
$$\bar{v} = \frac{s}{t} \quad \longleftarrow \text{distance travelled}$$

average velocity

And combining the above equations and rearranging gives

$$\boxed{s = v_1 t + \frac{at^2}{2}} \quad [= m] \quad \text{(ii)}$$

Now equations (i) and (ii) are not scientific laws but definition equations for uniformly accelerated motion. However, for freely falling bodies these expressions agreed with experiment, therefore they become physical laws for freely falling bodies as well. They are called Galileo's First and Second Laws.

### Newton's Three Laws of Dynamics and the Concept of Force

In kinematics we didn't ask, "Why does it move the way it does?". It is the object of dynamics to account for this action by saying that it is caused by something called a force. What is a force? It is not a thing - it has no shape, no mass, no color, no smell, and no smile. It is just an idea that we introduce in science which we find useful, such as energy and ether (see Chapter 12). We will introduce these and other magical and not real but useful quantities in later chapters. Let us see here how we use this concept of force.

### Newton's First Law of Motion  - A Qualitative Definition of Force

A body in motion continues moving forever in a straight line with a constant velocity unless acted upon by a force. A body at rest stays at rest unless acted upon by a force. So <u>an agent or thing that accelerates a body is a force.</u> This is called Newton's First Law.

### Newton's Second Law of Motion - a Quantitative Definition of Force

Force is a quantity proportional to the acceleration it produces on a body of mass, m, having the same direction as the acceleration. In the SI system the unit of force is called the Newton, N. So the second law of motion is: $\boxed{F = m \cdot a}$ ... with units: $\boxed{N = kg \cdot \dfrac{m}{s^2}}$  (iii)

where the unit of force is called the "newton" with symbol "N".
This law is a quantitative operational (or measurable) definition of the concept of force.

<u>Gravity</u>  Consider the earth and an object B. The force of attraction of object B to the earth's surface is called its weight, and is proportional to the mass of object B. Thus at earth's surface

$$F = m_B \cdot a_B = \text{Newton}$$

then becomes $\quad \text{weight}_B = m_B \cdot g \, [N]$

where the earth's acceleration constant at the earth's surface is

$$g = 9.8 \dfrac{m}{s^2} \quad [N]$$

**Newton's Third Law of Motion** - <u>Force on an Object</u>

Suppose a body of mass $m_A$ exerts for a time $t$ a constant force on mass $m_B$. Then for B

$$F = m_B a = \frac{m_B(v_2 - v_1)_B}{t} \quad \left[ = N = \frac{kg \cdot m}{s^2} \right]$$

Multiplying both sides by $t$ gives

$$Ft = m_B \Delta v_B \quad \overset{(v_2 - v_1)_B}{}$$

impulse to $B$ ↗ ↖ change in momentum

Or the <u>change in momentum</u> of a body <u>equals the impulse that causes it</u>

**Newton's Third Law** <u>for Two Colliding Objects</u>

If mass A collides with mass B then B reacts by exerting an equal and opposite force on A. In other words, to every action there is an equal and opposite reaction (as given by the minus sign in the equation following). As a consequence

$$F_A = -F_B$$

and if A and B are moving in opposite directions then if $v_A$ is positive then $v_B$ is negative.

To deal with situations where the forces change with time, Newton invented calculus. We don't deal with this here.

Let us now introduce the term "momentum" = $mv$. For two colliding masses, noting, that the time interval $(t_2 - t_1)$ during which the forces act on $A$ and $B$ are the same we end up with the momentum lost by one equals that gained by the other, or

$$\boxed{m_A \Delta v_A + m_B \Delta v_B = 0}$$

change in momentum of A ↗ ↖ of B

This is **Newton's Third Law** - or - <u>the Law of Conservation of Momentum</u>

Newton's three laws are the basis of what we call dynamics today, with its well known equations for falling objects or thrown baseballs which every freshman science and engineering student has met.

*Galileo and Newton started science as we know it today.*

## SUMMARY

The four basic equations from these two pioneers are, for two bodies A and B and with 1 = before, 2 = after the collision,

For object B
$$\begin{cases} v_2 = v_1 + at \\ v_2^2 - v_1^2 = 2as \end{cases}$$

For body B at constant acceleration the distance travelled
$$s = v_1 t + \frac{at^2}{2}$$

Force needed to accelerate body B
$$F = m_B a$$

For a collision of the two bodies which stick together

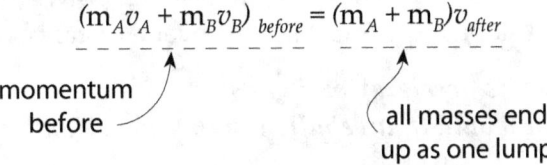

$$(m_A v_A + m_B v_B)_{before} = (m_A + m_B) v_{after}$$

momentum before

all masses end up as one lump

For all other types of collisions—elastic, angular, etc.— the analysis can become horrendous.

### PROBLEMS

Note: In all problems ignore air resistance. Also see Chapter 15 for conversion of units

1. A baseball pitcher can throw a baseball at 92 miles/hr. *If he throws the ball upwards how high in meters can it go?*

2. Ball A (3 *kg*) moving West at 10 *m/s* collides straight on with ball B (2 *kg*) moving due East at 5 *m/s*. *After collision where do the balls go and how fast do they go?*

    Balls A and B are perfectly inelastic and are of the same material. .

3. An anti-aircraft projectile is fired upward and reaches 40 000 ft.

   a) *For how long does it rise?*
   b) *For how long does it remain in the air?*
   c) *What is its initial velocity in m/s?*

4. A car moving at 15 mph is stopped in 5 seconds. Express its uniform acceleration in

   a) miles/hr · s    b) ft/s$^2$    c) m/s$^2$

5. A cube is released from rest and slides down a smooth inclined slope requiring 4 sec to cover a distance of 1 m.

   a) *What is its acceleration in m/s$^2$?*
   b) *How far would it have fallen in the same time?*

6. A car travels at 30 mi/hr for one hour and 20 mi/hr for 30 miles. How long does the whole trip take, and what is the average speed for the whole trip?

7. An empty freight car weighing 10 tons (metric) rolls at 1 *m/s* along a level track and collides with a loaded freight car weighing 20 tons, standing at rest with brakes released. If the two cars couple together, *what is their joint velocity after collision?*

8. In the previous problem with what velocity should the loaded car be rolling towards the empty one in order that both should be brought to rest by the collision?

PROBLEMS

9. An automobile travels at 60 km/hr for 30 km and then at 20 km/hr for 30 km.

   *What is the average speed of the car?*

10. A man is driving a car along a highway at a speed of 20 km/hr. He steps on the accelerator and one minute later the speedometer reads 80 km/hr.

    *What is the average acceleration of the car in ft/sec$^2$ ?*

11. A man at the top of a tall building throws a ball vertically downward at 20 m/s.

    *What is the downward displacement of the ball after 2 s?*

12. A 10 *kg* body experiences an acceleration of 20 *m/s$^2$* eastward.

    *What external force in newtons, acts on the body?*

13. If an elevator cage weighing 1000 *kg* experiences a downward acceleration of 4 *m/s*.

    *-What upward force in kg is exerted by the supporting cable?*

14. You are sitting and writing on a pad of paper in a comfortable, completely frictionless swivel chair and you want to turn clockwise 180°.

    *How would you do it without pushing on anything (such as a desk or the floor), fanning the air, or throwing things about the office?*

# CHAPTER 11

# THE STORY OF SCIENCE

*Science is to man as hot manure is to the pea.*
R.C. Punnett *Mendelism* 2nd ed. pg 8 1907

## THE PIONEER - GALILEO

There are many situations where Galileo questioned and challenged accepted orthodoxy. The most famous and important in the history of science concerns whether the earth was stationary (Aristotle's view) or whether it moves (Copernicus' suggestion). Galileo wrote an inflammatory book *Dialogue Concerning the Two Chief World Systems*, (1630), translated by S. Drake, Univ. of California Press, 1953, which started with the following message to the reader:

> "Several years ago there was published an order to impose a silence on the Pythagorean opinion that the Earth moves. Upon hearing such insolence and being thoroughly informed . . . I decided to appear openly . . . as a witness to the sober truth.
>
> To this end I have taken the Copernican side of the discourse . . . against the arguments of those who are content to adore the shadows, philosophizing not with due circumspection, but merely from having memorized a few ill-understood principles."

The Catholic Church was enraged, . . . how dare he! . . .

- They said that the notion of the moving Earth contradicted the Holy Scriptures.
- Father Riccoli enumerated 77 arguments against the Earth moving.
- The Church put Galileo's book on the Index, meaning that no practicing Catholic was allowed to read it.
- They put Galileo under house arrest for the rest of his life, which

- was a mild sentence and which terribly disappointed the official torturer.
- his book remained on the Index for 200 years, and finally, only in 1992, after a 14 year study, did the Pope and the church finally accept his ideas.

We will not go into this fascinating story. But regarding Galileo's new way of thinking which is the key to scientific thinking today Albert Einstein wrote:

*The dominant recurring theme which I recognize in Galileo's work is the passionate fight against any kind of dogma based on authority.*

Nowadays it is hard for us to grasp how sinister and revolutionary such an attitude appeared in Galileo's time when merely to doubt the truth of opinions which had no basis but authority was considered a capital crime and punished accordingly.

Actually we are by no means so far removed from such a situation even today as many of us would like to flatter ourselves; but in theory, at least, the principle of unbiased thought has won out.

## SCIENCE TODAY

Science, as a factor in human life, is exceedingly recent. Art and religion were well developed thousands of years ago, but science, as an important force in man's life began with Galileo and in Galileo's time about 350 years ago.

At first it was the leisurely pursuit of the learned and wealthy and it did not affect ordinary man. Only in the last 200 years did its applications develop the power of manipulating the world and change man's life. In its short time it has caused greater changes than in the thousands of years before that.

What is science? It is knowledge of a certain kind, that is arrived at by the **scientific method**. This involves

- collecting information by observation
- extracting general statements called laws from this information
- proposing hypotheses which connect, explain and account for these laws

It seems to be such a simple thing but in a magical way it brings us all the wonders of the marvelous technical world of today.

**Let us look at some examples of scientific laws and theories.**

## SCIENTIFIC LAWS

From the data of observations we discover general statements about classes of things. These are called 'laws', but other words such as 'rules' are sometimes used. It may not be obvious why a statement such as 'metals expand when heated' should be called a law.

In common speech a law is a command, and contains the idea of 'must' or 'ought'. Obviously 'thou shalt not steal' is in this sense a law. But no scientist is suggesting 'that a metal is ordered to expand on heating'.

The historical reason why the word law is applied is that the medieval theologians believed that the world is as it is because God has so willed it. Thus every true statement about nature is the result of His will or command - His law. But this is not what the scientist is talking about. By his law about metals expanding the scientist simply means that metals behave like that, and he certainly does not intend to imply that they do so because of any will, human or divine, or that any "intelligent design" is involved . It would be the place of theology or metaphysics, and not science, to discuss that question.

The simplest scientific laws are what we may call statements of properties. Thus:

- Ice melts at 0°C
- Gold has a density of 19 700 $kg/m^3$
- Salt crystallizes in cubes
- Cats have 20 claws

are all laws of science, statements about classes of objects; but we do not usually call them 'laws', a word we reserve for propositions which are more than mere descriptions, and which apply to a wide range of phenomena.

Laws which are statements about relationships between properties of classes are generally phrased in the form of algebraic equations.

Example: The increase in length (L) of a solid is directly proportional to the increase in temperature ($t$) through which it is heated, or

$$L_2 = L_1\{1 + a(t_2 - t_1)\},$$

where $a$ is a constant

Sometimes we can manipulate these equations and combine several laws, and we may sometimes predict new phenomena which were not foreseen from the separate laws before the mathematical treatment.

Example: Consider the following two generalizations or laws:

**Boyle's Law** states that for a given amount of gas at fixed temperature:

$$pV = \text{constant} \ldots \text{where } p = \text{pressure and } V = \text{volume} \quad (1)$$

**Charles' Law** (in England) or **Gay-Lussac's Law** (in France) states that for a given amount of gas in a fixed volume:

$$\frac{p}{T} = \text{constant} \ldots \text{where T is measured in K or °R} \ldots \quad (2)$$

Now without doing any more experiments, but by deduction alone, we can combine these two laws to obtain a more general law, applicable for any change to a given amount of gas

$$\frac{pV}{T} = \text{constant} \ldots \quad (3)$$

which tells us how any one of these three quantities p, V, or T will change when the other two quantities are altered. And as we obtained eq. (3) by deduction from eqs. (1) and (2) we know that it is as reliable as eqs. (1) and (2).*

This is a simple illustration of the advantages of phrasing laws as mathematical equations, because we can then perform on them mathematical or deductive operations and thereby obtain new laws. Equation (3) above is called the **Ideal Gas Law**.

---

\* *It's not so easy to derive eq. (3) from eqs. (1) and (2). Try it.*

## Scientific Hypotheses and Theories which have changed our World View

Laws summarize observations, while hypotheses and theories try to give an explanation to this and that law. If we would ask a medievalist for a typical explanation he may answer with:

Let a stone go and it falls; similarly a book falls, the apple that hit Newton on the head, falls. These are samples of the law of falling bodies. And the reason they fall is that God WANTS them to fall. It's as simple as that. That's the theory, the explanation for why they fall.

Another medievalist may give a different theory saying that all bodies want to go to their natural resting place, the center of the earth, which is the center of the universe. Let us illustrate these ideas with an example, one of the most important ideas in science.

## The Kinetic Theory of Gases

Man has speculated about the nature of things since time immemorial, however written records only go back as far as the Greek period. One of the questions about which the Greeks wondered was the following:

> "Could you take a piece of stuff and keep cutting it up indefinitely until nothing at all was left, or would you end up with an indivisible piece or grain of stuff?"

The word **atom** is derived from the Greek word meaning "indivisible", and this view that matter was ultimately atomic was held by Democritus and Epicurus. Now, no one then could check if this view was correct because not much was known about how things acted. They were right, but theirs were just lucky guesses.

The atomic theory was discredited for 2000 years because it seemed unreasonable: how could it explain color, odor, taste? Nevertheless it was brought up for discussion from the 17th century onwards due to newly acquired knowledge and because the other viewpoint could not explain how matter acted. After a number of tries the following explanation was attempted.

Let us imagine that a gas consists of trillion of trillions of very tiny particles of zero volume all darting about, here and there, all at the same unbelievably high speed of about 700 miles per hour. The particles collide with one another and with the container walls, bouncing off them like perfectly elastic billiard balls. It is this rain of the many, many, many particles pounding against the surface which is the ultimate cause of gas pressure. Also let us say that at higher temperatures the particles move faster (this is a shrewd guess based on Charles' Law). The skeptic would have some serious objections:

- If each particle has absolutely no volume then its density must be infinite because it does have some mass. This is crazy and doesn't make sense.

- How can the collisions be perfectly elastic? Certainly there would have to be friction, that's our world (if not we should be able to make a perpetual motion machine); and with friction eventually all the little particles would end up lying on the floor of the container, with a vacuum above. What a nonsensical idea!

- 700 mph! unbelievable!

Which explanation makes more sense, the one invoking God's will, or the deep desire of all things to approach their "home", the center of the universe, or the explanation dealing with stuff of zero volume zooming about at unbelievably high speed? What do you think?

If we put these objections aside and attempt to see what deductive consequences follow from this wild theory on atoms, those indivisible grains of stuff, we will be able to deduce:
    a) Boyle's Law,
    b) Charles' Law,
    c) the Ideal Gas Law,
    d) Graham's Law of Effusion,
    e) Avogadro's Law.

Qualitatively we will also be able to explain the mechanisms of:
    a) diffusion,
    b) heat conduction,
    c) gas pressure and
    d) viscosity.

What a powerful theory this is! — to be able to account for such a variety of facts and laws. <u>It is, in fact, one of the most powerful theories in physical science.</u>

Subsequent work went into refining this kinetic theory, but the core remains, the atomic nature of matter. This theory is refined by making changes in the assumptions

a) Atoms (or molecules) attract each other.
b) Atoms have a diameter different from zero.
c) Atomic velocities are given by the Boltzmann velocity distribution, rather than considering the velocities to be constant.

Each refinement allows us to approach reality more closely, however the mathematics becomes progressively more complicated. At some point we decide that it is not worthwhile to further complicate the picture. So utility somehow enters the picture. Each application has its appropriate model, and the choice is subject to economic considerations.

## THE THEORY OF EVOLUTION

Darwin's work is illustrative of the non-mathematical sciences. He dominated the intellectual outlook of his times, among the general public as well as the men of science.. He did not invent the hypothesis of evolution, which had occurred to many of his predecessors, but he brought a mass of evidence in its favor, and he invented a certain mechanism which he called **natural selection** to account for it.

Apart from scientific details, Darwin's importance lies in the fact that he caused biologists, and through them, the general public to abandon the former belief in the immutability of species, and to accept the view that all different kinds of animals had developed out of a common ancestry. Like every innovator of modern times, he had to combat the authority of Aristotle. Aristotle, it should be said, is one of the great misfortunes of the human race. To this day the teaching of philosophy in most universities is full of nonsense for which he is responsible.

The theory accepted by biologists before Darwin was that there was, up in Heaven, an ideal dog and an ideal cat, and so on; and actual dogs and cats are more or less imperfect copies of these celestial types. With this picture there could be no transition from one species to another, since each resulted from a separate act of creation. But geological evidence made this view increasingly difficult to maintain, so that today the theory of evolution, with its many corrections, is now practically universally adopted in our society, including grudgingly maybe even by the State of Kansas which, since 1996, has allowed it to be taught in its schools.

## RATIONALISM AND BUNKUM IN SCIENCE TODAY

Over 350 years ago rationalism was put in retreat by Galileo. Science and scientific theory was then born, and empiricism became its prime tool. However now and then concepts not grounded in observation and fact creep in and claim to represent reality (remember Aristotle's "nature abhors a vacuum") somewhat like a weed which intrudes into a neat tidy lawn.

Shermer (in the magazine called *Scientific American* pg. 24 Jan. 04) reminds us that the word BUNKUM means "empty claptrap oratory" and to DEBUNK means "to expose the nonsense and pretensions". He also states that:

> *no one talks bunkum with a better vocabulary than those who lace their bunkum with scientific jargon.*

In that very issue (pgs. 66 - 75) we see a proposed theory having a catchy title "Our Growing Breathing Universe". Quoting that article we read:

> *The spin networks that represent space in loop quantum gravity accommodate space-time by becoming what we call spin "foams". With addition of time the spin networks become 2 dimensional surfaces.*

Elsewhere the author adds time to the 4-d Einsteinian space-time to come up with a 2-d surface! That just takes my breath away. What does this crazy arithmetic mean? And again.

> *Time advances in ticks, but does not exist between ticks.*

and so on. These speculations remind me of the serious topics that concerned medieval scholars such as — the number of angels that could dance on the head of a pin.

To conclude, this proposed theory may be imaginative and interesting but it is not science. Or as the author finally admits in his conclusion:

*Everything I have discussed in this paper is theoretical*

This type of theory, as with Alice's "Adventures in Wonderland", belongs in the world of fiction, not science.

Or as Wolfgang Pauli would have stated
*it is not even wrong*

## Problems

1. **Intellectuals.** The Oxford English Dictionary defines 'intellectualism' as the 'doctrine that knowledge is wholly or mainly derived from pure reason'. *From what you have read in this chapter and in Chap. 1 would you say that:*

    (a) Aristotle
    (b) Galileo
    (c) Scientists

   *are intellectuals?*

2. **Medieval Italian Art.** Some artists drew Adam and Eve with smooth bellies having no navels because they were created by God, and were not born of woman. Other artists drew them with navels, saying that they were supposed to be created as perfect people, and all people, even perfect people, have navels. This difference in opinion led to enormous arguments.

*Which side would you back, the belly buttoners or the smooth belliers?*

3. **The Theory of the Spontaneous Generation of Life\*.** This theory played an important role in early biological studies. The appearance of living cells on plates where they had not been the previous day apparently happened so often that a detailed study of this phenomenon was not necessary.

\* Clifton, *"Introduction to Bacteria"* McGraw - Hill, New York (1950).

Around 600 B.C. philosophers of the Ionian school postulated that living things could originate from slime under the influence of heat, sun and air. Through Aristotle's studies (384-322 B.C.) the Christian church accepted this theory, and opposed free thinking on this subject. Much later Van Helmont, the famous alchemist and physician, proposed the following recipe for the spontaneous generation of mice:

> *"Place a dirty shirt in a vessel containing wheat and after twenty-one days' storage in a dark place, to allow fermentation to be completed, the vapors of the seeds and the germinating principle in human sweat contained in the dirty shirt will generate live mice".*

Even after the belief of the spontaneous origins of larger forms of life began to weaken, it continued strongly in relation to tiny living things. The Italian physician Redi in 1688 did away with the belief that decaying meat, in the presence of air and warmth, gave rise to small white worms. By placing a screen of muslin between the meat and flies, he showed that although air circulated freely and the meat decomposed, no worms appeared since they were nothing but fly larvae.

Others later showed that new organisms would not originate when materials were boiled for 15 minutes and stored in a sealed container. To counter this a Welsh priest, Needham, showed his faith in spontaneous generation by claiming that the vigorous heating destroyed the special vital force existing in every living thing and needed to form new living things.

Back and forth went the argument, and it wasn't until Pasteur sterilized broth in flasks with necks pulled out in the form of the letter S that spontaneous generation was finally disproved. The dust and germs in the air were trapped in the neck, and the broth was completely protected from contamination.

This simple but brilliant experiment won for Pasteur in 1862 a prize offered by the French Academy of Sciences for a convincing answer to the question of whether spontaneous generation did actually occur.

## WILD PROBLEMS

One of the best methods of getting practice in the entire process of scientific discovery and formulation (outside of long and intensive laboratory work), beginning with chaotic experiences is found in the solution of cryptograms.

In each of the following problems:

a. solve the cryptograms and state its key in the form of a rule or law, such as:
   - "Substitute for each letter the third following letter in the English alphabet"
   - "Replace each letter with the letter equally distant from the letter 'm'", etc.
   
   Some of the keys will be very simple, others rather complex.

b. Indicate the first few clues which helped you solve the problems.

In dealing with them you are somewhat like an archeologist attacking an inscription in an unknown tongue, or a Newton first disentangling the conception of mass, in terms of those ideas with which he proposes to describe the material world.

1.  SDRAWKCAB  DAER  TSUJ

2.  BCE  DOAF  BO  CEGG  IB  JAKEF  LIBC  MOOF  INBENBIONH

3.  ZMHDVIH  ZIV  GL  YV  TFVHHVW,  MLG  TREVM
    —Bilbo to Gollum in *The Hobbit*

4.  88O  IE  O77  6O88  88O  1E  8833A88  I8  8833E  7UE888IO6
    —Shakespeare

5.  KE  FKO  QOYREQ  DKIQ  QOYREQ  TUHK

6.  WLVH  GSV  NLIGZI  SLOW  GSV  YIRXPH  GLTVGSVI  LI  PVVK  GSVN  ZKZIG?

7.  MDUDQ  KZTFG  ZS  KHUD  ZMC  GTMFQX  CQZFNMR -
    Bilbo's favorite saying

CHAPTER **12**

# THE FIRST LAW AND THE CONCEPT OF ENERGY

*Engineering aims to harness the powers and forces of nature for the benefit of man.*

For over 1000 years, from ancient Greek times, people shopped in a marketplace of ideas about our world. Everyone had his own opinion about how and why things behaved as they did - where the center of the world was, why things fell, what caused them to move, where your feelings came from, and so on. This was a **rationalist** approach to knowledge. For example do you want to convince your husband to buy you a new dress? Prepare a delicious meal for him, that's the key, because his stomach is where his feelings are. Do you want it known to the world that you love her? If so then carve a heart on a tree trunk, and so on.

However around 1200 AD the particular ideas of Aristotle and Plato were adopted by the Catholic Church and were made part of their dogma. They asserted that the opinions of those ancients were the truth and if you disagreed then you were a heretic and had to be warmed up a bit. At this point the rationalist approach was transformed into an **authoritarian** approach to knowledge.

Galileo said "no", you learn about our world by observing, measuring and then reasoning based on these observations. This is the **empirical** method, an absolutely revolutionary approach to knowledge in Galileo's time. From that time on scholars started relating all sorts of phenomena, for example:

- velocity of a falling object with distance fallen (Galileo)
- heat generated with work done (Benjamin Thompson, Meyer, Joule)
- heat generated by a given flow of electricity (Faraday)

- heat released during a chemical reaction occurring at constant volume (Helmholtz), and at constant pressure (Gibbs)
- heat generated by the disappearance of matter (Einstein)

In 1804 Thomas Young came up with an new word and concept called "energy". What was it? It was not a thing, - it had no mass or volume, no size or shape, no color or smell, and you could not touch it or experience it in any way. It was an abstract idea which was difficult to understand and to make sense of.

In 1842 Meyer crystallized this vague idea by saying that when change occurred by any of the following phenomena:

- food digestion - heat generation - electricity flow - a burning fire - a speeding body - work done

this magical thing called energy was the thing that was conserved. Thus when one thing lost some energy, something else gained an equal amount of this magical stuff. It was the concept of energy that tied together these many different phenomena.

When Meyer tried to publish his ideas on energy the journal editors considered this to be nonsense, and they rejected his paper. The next year Helmholtz's paper on the same subject suffered the same fate.

This idea took quite a while to become accepted because it opposed the prevailing view at the time that living things had a "vital force" that made them live. That idea died hard. As recently as the 1930's Soviet scientists very carefully weighed people just before and just after they died to see if they could find the weight of a person's soul. Even today chiropractic practice is based on the conviction that a "life force" or a "life energy" exists that moves up and down a person's spine.

Today we consider energy to be one of the central concepts in all of science. We define the First Law of Thermodynamics as follows:

**Energy cannot be created. It cannot be destroyed.**
**It can only be changed from one form to another.**

The basic measure of energy $(E)$, chosen by the world's scientific community in the 1960s is the **joule** $(J)$. This is defined as the force of one newton $(N)$, pushing something forward one meter $(m)$. Thus:

$$1 J = (1 \text{ newton})(\text{meter}) = 1 \, N \cdot m$$

## Summary of Mechanics    12.3

As a rough idea of how big a joule is

- it takes 3 J to lift a can of Coke from the floor to a table (potential energy)
- it takes 140 000 J to bring a can of Coke from the refrigerator to a boil (thermal energy)
- it takes 4 J to heat a thimbleful of water 1°C (thermal energy)

Here's how we measure the energy increase $\Delta E = E_2 - E_1$ of a mass m for given changes

- the mass is raised from $z_1$ to $z_2$

$$\Delta E = mg(z_2 - z_1) = kg \cdot \frac{m}{s^2} \cdot m = N \cdot m = J$$

$$\text{note}: N = \frac{kg \cdot m}{s^2}$$

- the mass is speeded from $v_1$ to $v_2$

$$\Delta E = \frac{m(v_2^2 - v_1^2)}{2} = kg \cdot \frac{m^2}{s^2} = N \cdot m = J$$

- the mass is heated from $t_1$ to $t_2$

$$\Delta E = Q_{added} = mC_p(t_2 - t_1) = kg \cdot \frac{J}{kg \cdot K} \cdot K = J$$

where  $Q$ = heat added = J

$C_p$ = specific heat of the object = $\frac{J}{kg \cdot K}$

= heat needed to raise the temperature of 1 kg of the object 1 K

= 4184 $\frac{J}{kg \cdot K}$  ... for water

= ~1 $\frac{J}{kg \cdot K}$  ... for air

$\lambda$ = heat needed to boil water = 2260 $kJ/kg$

= heat needed to melt ice = 234 $kJ/kg$

- for an electric current
$$\Delta E = V \cdot A = (volt)(ampere) = [\frac{J}{s \cdot A} \cdot s]A = J$$

- when 1 mol of **A** at temperature T and pressure $p$ is transformed by chemical reaction into products **R** and **S** which end up at the starting T and $p$ by the reaction

$$A \longrightarrow rR + sS, \ldots \Delta H_r$$

where $\Delta H_r$ is called the enthalpy of reaction, per mol of A reacted

if $\Delta H_r > 0$... heat released by the reacting mixture

if $\Delta H_r < 0$ ... heat added to the reaction mixture.

- if mass disappears, it is transformed into energy $\Delta E$ according to Einstein's famous equation

$$\Delta E = -\Delta M \cdot c^2 = kg \cdot \frac{m^2}{s^2} = N \cdot m = J \qquad \boxed{note : N = \frac{kg \cdot m}{s^2}}$$

where M = mass, kg
  $c$ = speed of light = $3 \times 10^8 \; \frac{m}{s}$
  J = unit of energy = $\frac{kg \cdot m^2}{s^2}$

The joule is the standard unit of energy, however there are various other ways of measuring it. Here is a table of equivalence to one million joules, from very small to very large.*

$10^6$ J = 6.24 x$10^{24}$ eV (electron volts)—used in elementary particle physics

  = $10^{13}$ ergs—the old standard metric measure, the cgs system

  = $10^6$J - today's standard, the SI measure

  = 737 562 ft · lb$_f$ (foot pound force)—an English system unit

  ≅ $10^{-9}$ **QJ**

* "For those who want some proof that physicists are human, the proof is in the idiocy of all the different units which they use for measuring energy."
    —"The Character of Physical Law" by Richard Feynman

## Various Measures of Energy

$10^6$ J = 2.39 × $10^5$ cal (calories)—old metric system, or 4.184 J = 1 calorie

= 1.02 × $10^5$ $kg_f$ · m—old metric system

= 9.87 × $10^4$ L · atm (liter · atmosphere)—old metric system

= 947.8 Btu (British thermal unit) — a commonly used English system unit

= 239 kcal or Cal—(a big calorie or a food calorie)

= 0.372 5 Hp · hr (horsepower · hour)—English system unit

= 0.278 kW · hr (kilowatt · hour)—often used in practice

= 27.2 × $10^{-3}$ $m^3$ of natural gas—a gas industry measure

= 9.478 × $10^{-3}$ therm—a gas industry measure

= 2.39 × $10^{-5}$ tons of oil—a petroleum industry measure

= 1.111 × $10^{-11}$ kg of mass—(from E = $mc^2$)

1 QJ = 1.00 × $10^{15}$ Btu ≅ 1.055 × $10^{18}$ J — quad

The production rate of doing work (the power) is the energy produced in unit of time. In the SI system the standard measure is the watt, W, or

$$1 \text{ W} = 1 \text{ J/s} = 1 \text{ N·m/s}$$

The conversion of units is as follows

1000 W = 1 kW = 1341 Hp

### Quantities related to a Watt

| | |
|---|---|
| nano W = $10^{-9}$ W | kilo W = 1000 W = 1 kW |
| micro W = $10^{-6}$ W | mega W = $10^6$ W = MW |
| milli W = $10^{-3}$ W | giga W = $10^9$ W = GW |
| | Q W = 1.055 × $10^{18}$ W |

### Rate of Energy Use of Power Producers (J/s or W)

pumping human heart   3 W = 3 J/s

bright light bulb   100 W

working man (long term)   20 W
   (short term, 1 hr)   100 W

working horse   1 kW = 1000 W

lady's hair dryer   1 kW

compact automobile  (maximum available)  100 kW = $10^5$ W

Boeing 747 (at take off)   200 000 kW

large coal fired electrical power plant   1 GW = $10^9$ W

space shuttle orbiter with boosters (at take off) 14 GW

US total energy use 100 **QJ/s** $\cong$ 1 x $10^{20}$ W

man's total rate of energy use in the world today $\cong$ 400 **QJ/s**
$\cong$ 4 x $10^{20}$ W

## THE LATEST SAGA OF MAGICAL QUANTITIES IN SCIENCE

The ancients and the medievalists built their world of mechanics and physics on a couple of principles which we mentioned in Chapter 1, that:
- nature abhors a vacuum
- only one force at a time can act on a body

Somewhat similarly, modern physics since Galileo had its abstract ideas. The idea of "energy" was one of these. Another is the "field" (see chapter 14). And still another was that magical thing called the "ether". *Let me try to explain.*

Sound waves travel through air, but if you start to evacuate a sealed bell jar which contains a ringing bell, the bell will sound quieter and quieter as air is being removed. When all the air is gone you won't hear the ring at all, even though you will see the clapper hitting the bell. So waves travel in something, some sort of medium.

Similarly, light travels as waves, so what sort of medium does it travel in? It has to be present in a vacuum, in air, in glass, in your eye because light travels through these media. Since we do not know what it is, we give it a name, the **ether**. Here is what the dictionary says about the word "ether".

# Magical Quantities—Ether, Energy

*It is a medium of unusual qualities. Although it lacks stuff and substance it has absolute continuity, high rigidity and elasticity. It is the medium that transmits light waves and radio waves.*

But how to measure how fast it is moving past us? Here is a possible way; consider this analogy.

If you row a boat downstream in a river you will go faster than if you row across the river or upstream against the current. So if you measure the time going one way, and then the other, you should be able to calculate the speed of the river flow. Similarly, if you measure the speed of light when the earth is going one way and then the other way through the ether (a difference of at most 60 km/s (or 36 miles/s), see the sketch below), you should be able to determine the ether wind, or how fast our earth and our solar system are flowing through the universe and the ether. *Now that is really exciting!*

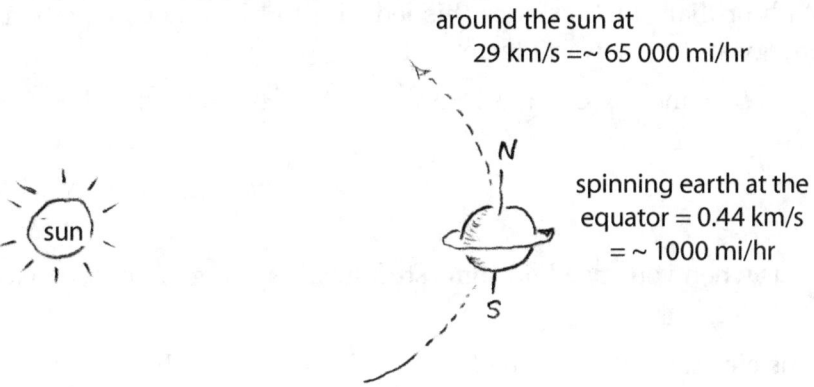

A physicist, Michelson, and a chemist, Morley, from U.S. Midwestern universities, prepared and did this most exciting physics experiment of the 19th century. In 1887, the whole world of physics awaited excitedly for the results. What did they find? To their puzzlement and disappointment no ether wind was observed!

This was unexpected. It just couldn't be, and it caused much consternation in the world of physics. All sorts of explanations were suggested including that matter expanded when moving in the direction of the ether flow, and shrank when moving in the opposite direction. But it was to no avail. It seemed that nature was in a conspiracy to prevent our finding how our solar system flowed through the universe, or that God did not want man to learn his secrets.

But then, about 18 years later, in 1905, a young 25 year old physicist working as a examiner in the Swiss Patent Office said

*"If theory and experiment give contradictory results I will do as Galileo did, I will accept what experiment says is so (here is the empiricist talking), even though it means that much of our fundamental concepts in physics has to be thrown out."*

Why didn't any other physicist have the courage to believe what experiment told him. Why did they prefer to accept without question what theory said and not experiment. It seemed that rationalism was again raising its head. Anyway, Einstein said

*"I will believe what experiment said, that light travels at the same speed in any and all directions, and see where it leads me."*

With brilliant mathematics this led him to his famous equation, that energy and mass was interrelated

$$E = mc^2 \ldots \text{or} \ldots \Delta E = \Delta mc^2 = kg \cdot \frac{m^2}{s^2} = N \cdot m = J$$

*and again* **note**: $N = \frac{kg \cdot m}{s^2}$

Thus when you speed up some stuff it gains energy and it increases in mass.

Einstein not only modified Newton's laws with these findings, but altered our conception of the universe. Scientists are agreed that Einstein was a genius the like of which may not appear again in generations. He has been hailed as the greatest intellect of our age. And what did he do for thermo? He accepted the findings of experiment instead of trusting what accepted reason said had to be true. He was a modern day Galileo, a true empiricist, not a rationalist.

However some people stubbornly stick to old ideas. For example a quarter of a century after Einstein published his work on Relativity, a book was published in Germany called *100 Authors against Einstein* which sought to show that Einstein must be wrong because so many opinions were ranged against him. See:

*A Random Walk in Science*, pg. 90, or *Die Naturwissenschaften*, **11**, 254 (1931)

Let us look at thermo today. We have energy equivalences for the whole range of phenomena:

- Galileo tied potential energy (gravitation) to kinetic energy.
- Joule brought in heat and work and friction.
- Faraday brought in electricity and magnetism.
- Helmholtz and Gibbs brought in chemical reactions,
- And in the final step, Einstein completed the map of thermodynamics and energy, by relating matter to the other forms of energy

Graphically we show this overall picture as

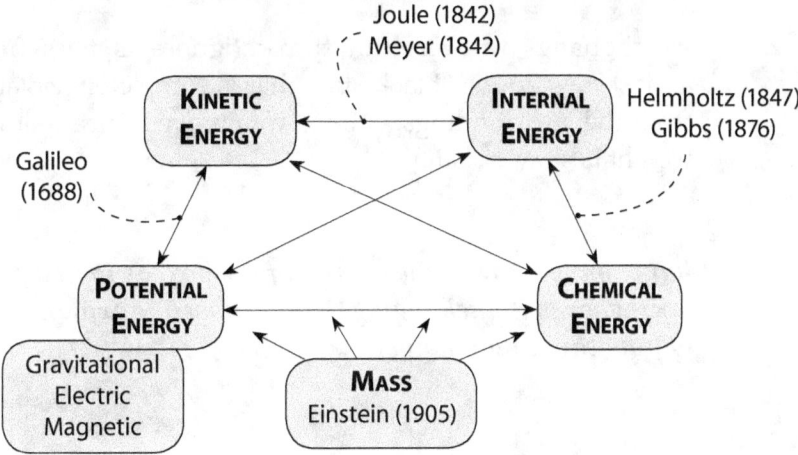

## FINAL REMARKS

The First Law relates a whole host of phenomena in terms of energy.

However there are some scholars who propose that there are still other forms of energy, such as psychic energy, life energy or 'vis viva', mental energy, athletic energy, and so on. However those terms and concepts do not fit into the framework which was developed in this chapter. They are not part of thermo.

So unless we can come up with some other form of energy which can be tied to those dealt with here, it seems that with Einstein's creation we now have a complete picture of the concept of energy as it ties together the various concepts in physics.

**The First Law of Thermodynamics is the central underlying concept in physics today.**

## Problems

1. The door to an ordinary electric home refrigerator is left open by accident (with the power on) while the owners are away for the weekend.

   *If the kitchen doors are closed and the kitchen is thermally well insulated will the kitchen be hotter or colder or at the same temperature as it was on Friday when the unhappy owners return? Why?*

2. A weight hangs by a very thin thread (ignore its mass) from a most mysterious "black" box that is completely isolated from the surroundings. As we watch, we notice that the weight is slowly rising.

   *What is happening to the energy of the box? Is it rising, falling, or is it unchanging? Give a possible explanation of how this could happen.*

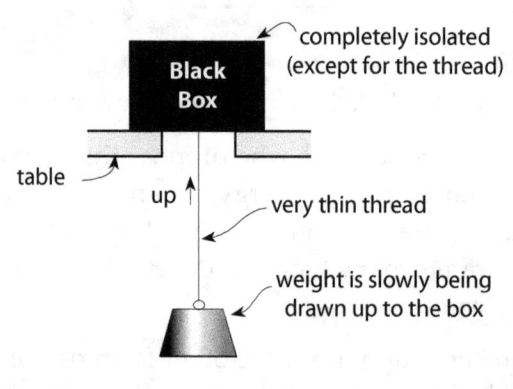

3. The sun radiates at a rate of $3.9 \times 10^{24}$ kW.

   *How many metric tons of mass is lost by the sun each day?*

4. From *The Feynman Lectures in Physics* Vol. 1, pg 45-4, California Institute of Technology (1963) we read:

   *". . . let us consider a rubber band. When we stretch a rubber band we find that its temperature falls . . . "*

   a) *Do an experiment to check this statement. Take a rubber band, pull and release. While touching the band to just above your upper lip, decide whether you agree with the above quote.*

   b) *Then try to explain your finding in terms of thermo.*

5. Squeezable but incompressible Bubbles-la-Rue is innocently floating in the swimming pool when Zoran-the-Mean, who is skulking by the side of the pool, reaches over with a stick and pushes and holds her under water. Naturally our dear Bubbles is frozen with surprise.

   *From the thermodynamic point of view, how has his dastardly act affected:*
   a) *her energy (up, down, or unchanged)?*
   b) *the energy of the water?*

6. A powerful electric fan is switched on in a closed insulated room causing the air to circulate in a clockwise direction. After 14 hours the fan is reversed and the air circulates in the counterclockwise direction. After 28 hours the fan is switched off.

   *What can you say about the temperature and energy of the room before and after the 28 hours?*

7. A healthy diet for a normally active person should supply 2000 Cals/day.

   *If you and your backpack (plus clothes, of course) weigh 100 kg, how high could you climb on one day's food intake assuming complete energy conversion?**

   * Note that 1 big calorie = 1 Cal = 1 kcal = 1000 cal = 4184 J

8. The ocean liner, Queen Elizabeth, has a mass of 85 000 tons and it travels at a top speed of 28 knots.

   *What is the increase in mass of this liner (in kg) when moving at its top speed when compared to its rest mass?*

   Data:  1 knot = 1 nautical mile/ hr
   1 nautical mile = 6080 ft

   *also see Chapter 15*

9. What will happen to the temperature of a thermally insulated room if you plug in

   a) an electric fan?

   b) a radio?

CHAPTER 13

# THE AMAZING SECOND LAW

*Engineering is the art of applying science for the benefit of mankind.*

## THE BIRTH OF HEAT ENGINES

Until the end or the sixteen hundreds man had to use the energy of his muscles, those of his slaves, of his horses and of flowing water and of the blowing wind to do useful work for him. But in 1698 Thomas Slavery, for the first time in man's history, was able to get useful work from the flow of heat from steam to cold water. His invention was a great boon to the mining industry and was used to pump water up and out of coal mines. The reason was that water often leaked into and flooded coal mines, and either had to be removed by a bucket brigade or else the mines had to be abandoned, which was a disastrous situation.

His device was crude and it had no moving parts. Here is how it worked:

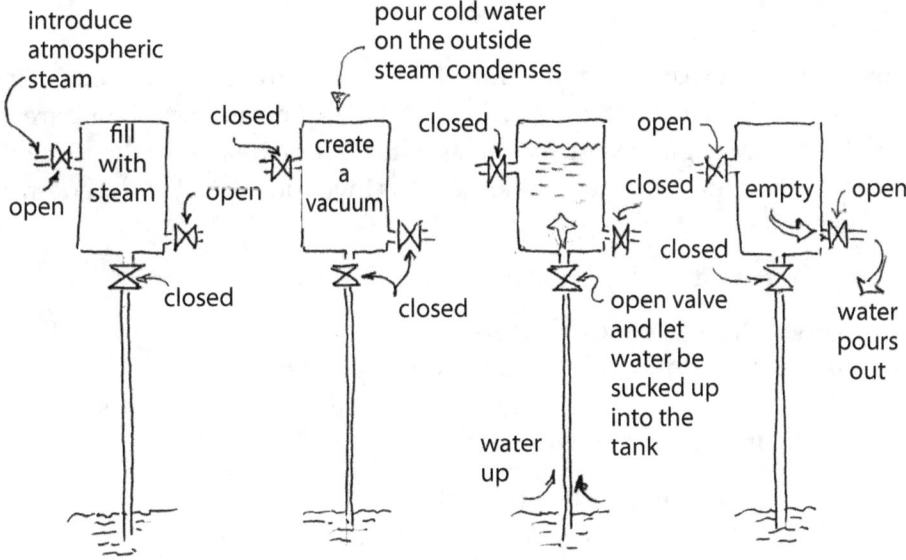

This was called the steam pumping engine. Slavery also coined the term horsepower Hp.

*Advances quickly followed*

Newcomen built a better engine, open at the top and with a moving piston, as shown below:

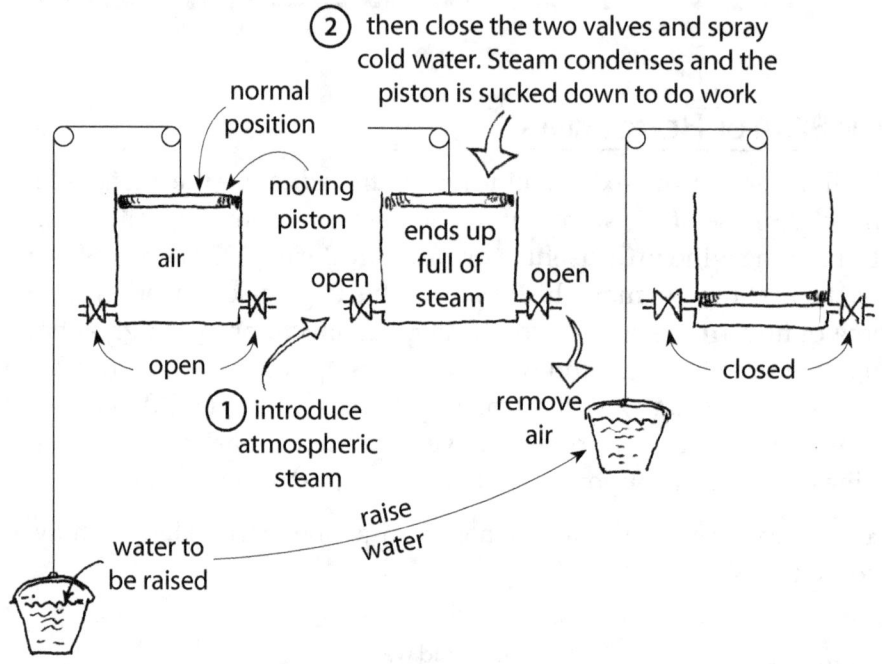

By 1769 the Newcomen engine ran at 5 strokes/min and produced 5 Hp of work. All you had to do was to turn the valves on and off. Humphrey Potter, a lazy engine boy used strings and levers to operate the valves. He was fired for sleeping at his job, but his invention doubled the engine speed to 10 strokes/min.

James Watt made many more improvements. He

- closed the top of the cylinders
- introduced steam at both ends of the cylinder
- did not use cold water, but discarded the waste steam, thus giving us the modern steam engine.

Today our modern societies run on this concept of heat flow from a high temperature $T_A$ (for example burning gasoline), to low temperature $T_B$ (exhaust), to produce work. Think of it, what would life be like

# A Young Sadi Carnot's Genius

without automobiles, coal fired electric power plants, jet planes, and your power lawn mowers. They all operate on the principle which followed the dream of Thomas Slavery.

## Sadi Carnot's Creation

When these engines were being invented and developed - the gasoline and the steam engine - the type of question that nagged the engine builders was; how much work could you get from a locomotive for each ton of coal used? How much water could you pump up from a flooded mine for each kilogram of coal used, how far could a car travel on one gallon of gasoline. Today we read in the papers that some engine designers talk of getting 1000 miles/gal of gasoline in the auto of the future.

A few years ago an inventor in Bend OR developed a device to be connected to an electric automobile. It had a big propeller and when you drove the car the propeller spun and generated electricity which was transmitted to the car's electric motor.

When the newsman asked if it really generated enough electricity to run the car, the inventor replied "Of course it will when the car goes at over 50 mph; I just haven't gone that fast yet, but I will soon. As evidence of the soundness of my invention I have a group of entrepreneur - businessmen queueing up to get a franchise to sell this revolutionary machine".

Sadi Carnot a young French engineer asked this question facing engine designers in the following manner

*"What fraction of the heat flowing from hot $T_A$ to cool $T_B$ could be transformed into work"*

or graphically

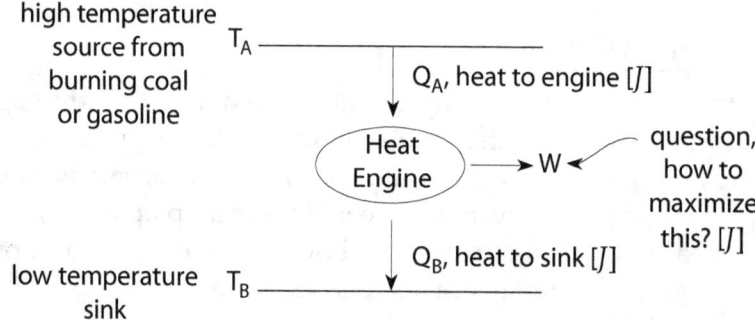

Starting with the first law and with the observation that heat wouldn't, on its own, flow "uphill", with brilliant but simple reasoning (no mathematics) Carnot came to the following conclusions (see the last part of this chapter).

1. Ideally there is an engine which can be run forwards or backwards. This unique engine is called the **Reversible Engine**.

2. The reversible engine is the most efficient of engines, in that it will produce the most work for a given heat flow between any two temperatures. Today we call this

3. **The Carnot Heat Engine**

The maximum extractible work from the ideal engine is given by

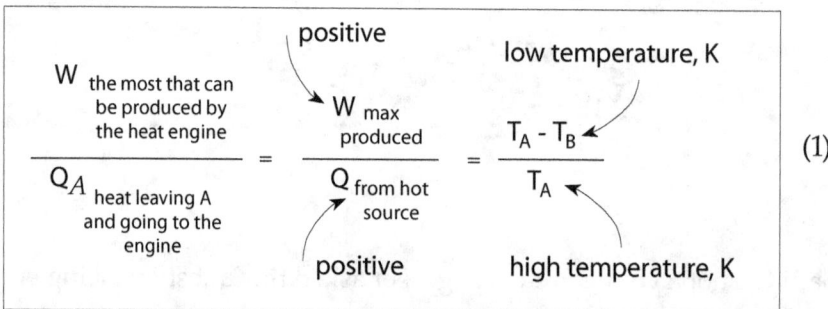

(1)

where the temperature is measured in absolute (or Kelvin) units.

# The Best Heat Engine

In getting to this equation Carnot invented the concept of **entropy**, led the way to the Kelvin temperature scale, and came up with what we today call the 2nd law of thermodynamics.

4. When run backwards the engine is called the **Carnot Heat Pump**. This is the ideal for today's refrigerators, freezers, auto air conditioners, etc. in that it requires the least amount of work to do a particular amount of cooling

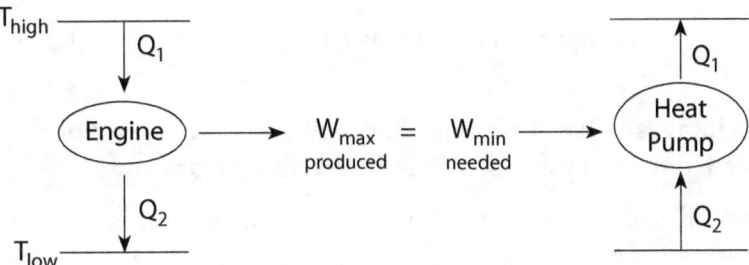

The minimum amount of work needed to remove one unit of heat from the cold region is given by

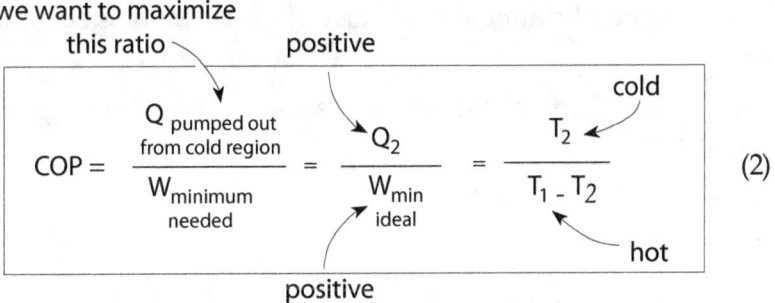

$$\text{COP} = \frac{Q_{\text{pumped out from cold region}}}{W_{\text{minimum needed}}} = \frac{Q_2}{W_{\text{min ideal}}} = \frac{T_2}{T_1 - T_2} \qquad (2)$$

The <u>coefficient of performance</u> is called the COP.

## The Arrow of Time

The first law says that whenever a system and surroundings exchange energy the total energy involved in the exchange remains constant. So if one object loses energy the other object gains that energy.

Our experience tells us that things left to themselves change one way with time. They do not go one way today, the other way tomorrow. For example:

- Hold a rock. Let it go and it falls. It does not rise from floor to your hand.
- Black coffee plus cream gives 'white' coffee.
- Mix hot water and cold water and you get warm water.
- A piece of cotton wool soaked in alcohol plus a lighted match burns.
- Shuffle an ordered deck of cards (A, 2, 3,...) and you get a shuffled deck.

These changes will not reverse themselves and go the other way. The final state of these systems is called the **equilibrium state**.

The second law law accounts for this phenomenon. It is the 'arrow of time'. It tells which way things will change, it tells the way to equilibrium.

In some situations it may be difficult to tell where equilibrium is, so thermo has developed a measure to help us find a way. This measure is called the entropy, $S$, and the second law says that as a system approaches equilibrium the total entropy of system plus surroundings rises.

At equilibrium the total entropy of system and surroundings becomes maximum.

## MEASURING ENTROPY CHANGES

Quantitatively, the change in entropy of an object as it goes from state 1 to state 2 is given by

$$\Delta S_{object} = \frac{Q_{\text{added to object when the heat is added most efficiently, without friction}}}{T_{\text{of object during heat addition, as it goes from } T_1 \text{ to } T_2}} = \int_{T_1}^{T_2} \frac{dQ}{T} \qquad (3)$$

## IMPORTANCE OF THE SECOND LAW IN ENGINEERING

If left alone a system will approach equilibrium and the total entropy of system and surroundings will increase. But with the aid of a clever engineer some work can be extracted from the system as it approaches equilibrium. But as he does this the entropy increase is smaller. In the extreme situation **the second law tells what is the maximum amount of work that can ever be produced, even by the cleverest of engineers. And this is all done with no increase in entropy of system plus surroundings.**

This is the prime use of of the 2nd law; to tell you what is the maximum amount of work you can get out of a system

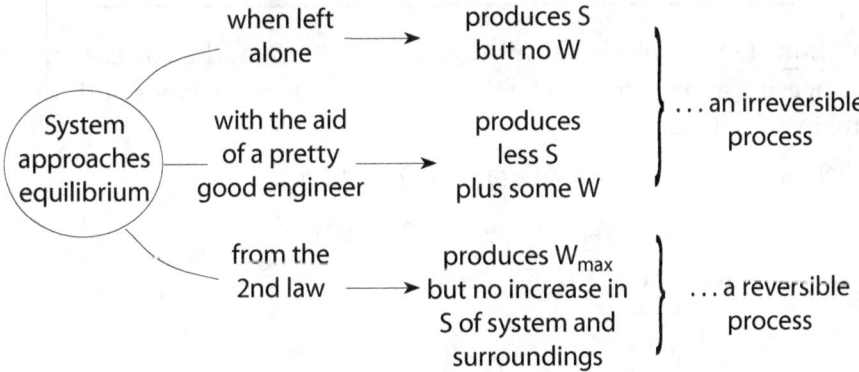

## MEASURING ENTROPY CHANGES AND WORK GENERATED

The following examples show how we link entropy to work and how we calculate entropy changes. But with the aid of a clever engineer some work can be extracted from the system as it approaches equilibrium.

**Example 1.** The Calculation of Entropy changes with Temperature

A block of hot concrete ($m_A = 1\,kg$, $T = 127°C$, $C_p = \dfrac{800\,J}{kg \cdot K}$) is dropped into a swimming pool of water ($T = 27°C$, $C_p = \dfrac{4184\,J}{kg \cdot K}$) What is the entropy change

(a) of the concrete block, $\Delta S_A$?
(b) of the water, $\Delta S_B$?
(c) overall?

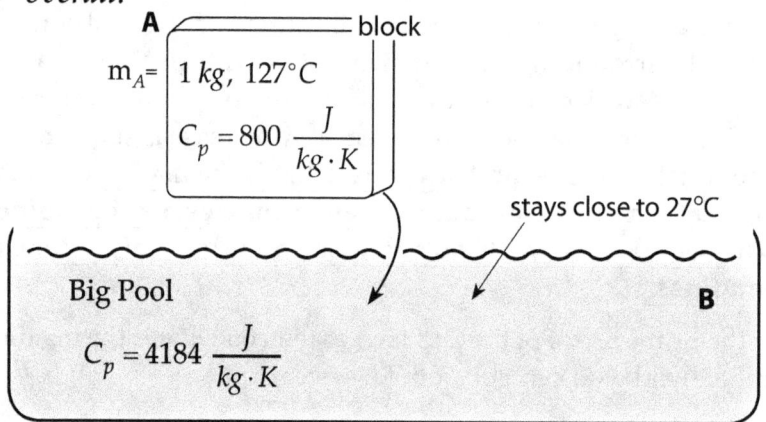

**Solution** Let the block be called **A**, and water **B**, and as a reasonable engineering approximation let the final temperature of block and water remain close to 27°C.

a) Then the heat added to **A**: $Q_A = m_A C_p (T_2 - T_1)$

$$= (1)(800)(300 - 400) = -80\,000\,J$$

Entropy change of A from eq. 3:

$$\Delta S_A = \int_{T_1}^{T_2} \frac{dQ_A}{T_A} = m_A C_p \int_{400}^{300} \frac{dT_A}{T_A} = 800\,\ln\frac{300}{400} = -230\,\frac{J}{K} \quad \text{← entropy lost}$$

b) From the first law the heat added to **B**: $Q_B = +80\,000\,J$

So entropy change of **B**: $\Delta S_B = \dfrac{Q_B}{T_B} = \dfrac{80000}{300} = 267\,\dfrac{J}{K}$ ← entropy gained

## Tricky Examples

c) For the whole 'universe': $Q_{A+B} = -80\,000 + 80\,000 = 0$

So $\quad \Delta S_{total} = \Delta S_A + \Delta S_B = -230 + 267 = +37\,\dfrac{J}{K}$, and this is $> 0$

**Example 2.** <u>Find $W_{max}$ available when heat flows from hot **A** to cool **B**</u>
800 J of heat flows from source **A** which stays at 1000 K to sink **B** which stays at 400 K, and on the way does work. What is $W_{max}$, the most work that can be produced, Also, what is $\Delta S_A$, $\Delta S_B$, and $\Delta S_{total}$?

```
                                              T_A = 1000 K
Gained by source ... Q_A = -800 J      ↓
                                    (engine) → W_max produced
Gained by sink ... Q_B discarded       ↓
                                              T_B = 400 K
```

**Solution** For source **A** heat is lost so the heat gain is:

$Q_A = -800\,J \ldots$ Thus $\quad \Delta S_A = \dfrac{Q_A}{T_A} = \dfrac{-800}{1000} = -0.8\,\dfrac{J}{K}$

From eq. 1, noting that $T_A = 1000$ K and $T_B = 400$ K the work produced is

$$W_{max} = \dfrac{T_A - T_B}{T_A} Q_A = \left(\dfrac{1000 - 400}{1000}\right) 800 = 480\,\dfrac{J}{K}$$

Since the heat lost by **A** equals the work produced plus the heat rejected to **B**, we have in symbols

$\qquad -Q_A = W_{max} + Q_B$

or $\qquad 800 = 480 + Q_B$

or $\qquad Q_B = +320$

Therefore $\quad \Delta S_B = \dfrac{Q_B}{T_B} = \dfrac{+320}{400} = 0.8\,\dfrac{J}{K}$

and $\qquad \boxed{\Delta S_{total} = -0.8 + 0.8 = 0}$

*Note:* Part of a system can lose entropy, but if the whole system and surroundings is considered their total entropy can never decrease.

**Example 3.** The Ideal Home Refrigerator

Heat has to be pumped out from the cold box of an ideal refrigerator which is at $T_B = -15°C$ and released into the kitchen which is at $T_A = 25°C$.

*We want to know how much heat can be removed from the cold box for each unit of work done.*

**Solution.** Equation 2 says that the amount of heat pumped from the cold box for each unit of work done on the system, in the ideal situation, is

$$\frac{Q_B}{W} = \frac{Q_{removed}}{W_{min}} = \frac{258}{298-258} = 6.45 \frac{J_{of\ heat\ removed}}{J_{of\ work\ done}}$$

This result is shown in the sketch below

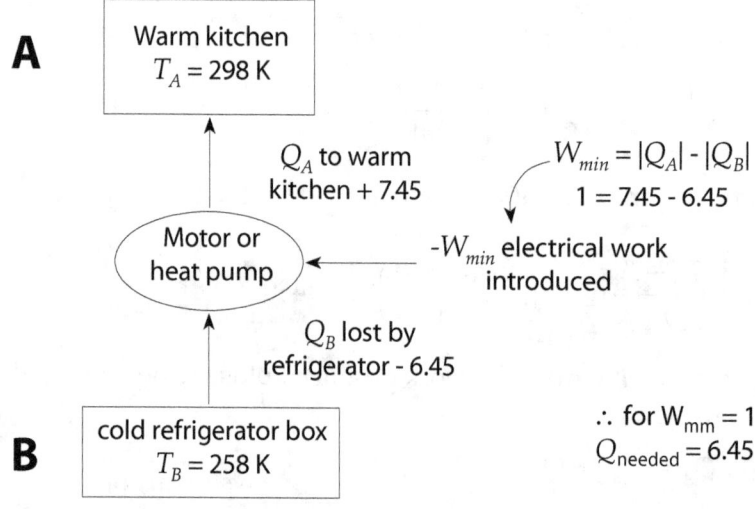

### Summary of the 2nd Law for Systems dealing with Heat, Temperature and Work

To help visualize the ideas of the 2nd law consider the flow of hot fluid at temperature $T_A$ through a clever device called an engine which is designed to produce work and then rejects fluid at a cooler temperature $T_2$.

1. It defines an abstract term called ENTROPY, S, (somewhat as we defined the abstract term ENERGY, E), and tells how to measure its change when the hot fluid goes through the engine.

2. When the flowing fluid tries to do work in practice, the total entropy change for system plus surroundings increases, it never decreases.

3. When the engine becomes more efficient it produces more work The total entropy change is still positive but becomes smaller.

4. In the ideal case when you are able to extract the most possible work from the hot-to-cold fluid, then you have $W_{max}$ and $\Delta S_{total} = 0$ These operations are called REVERSIBLE.

## THE MANY OTHER ASPECTS AND APPLICATIONS OF THE 2ND LAW

So far we have only considered the engine that extracted work when hot fluid flowed from a high to a low temperature. This arguably was the greatest technological invention of all time and was the key to the development of the industrial societies of our times. Also, we could point out that the word engineer originally referred to the person who tended these engines.

However the 2nd law is much more general and is applicable to all sorts of other phenomena and changes, such as

1. Exergy and availability.
2. Chemical reactions and free energy.
3. Mixing and separation.
4. Entropy and information theory.

*Let us touch lightly on some of these topics.*

### 1. Exergy and Availability

When a fluid goes from state 1 (say at 500K) to state 2 (say at 400K) while the surroundings are at some other state (say at 300K), the 2nd law says that the most work that can be produced is the Carnot work for heat flowing from 500K to 400K, (see eq. 1 and $W_{1 \to 2}$ in the next figure).

But extra work can be produced if the heat is rejected at the surrounding conditions at 300K

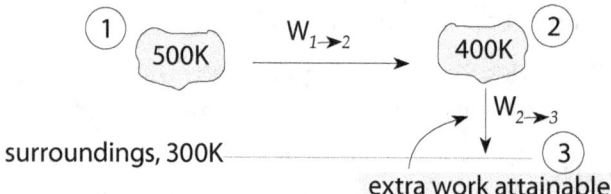

surroundings, 300K

extra work attainable

The total work, including the extra term is called the **availability** (in the US) or the **exergy** (in Europe and Japan)

The exergy is then the total work that can be obtained when fluid flows from $p_1$ $T_1$ to $p_2$ $T_2$ and then to surroundings at $p_0$ $T_0$. The work is calculated from $p$ $V$ $T$ and $S$ for states 1 and 2, plus $p_0$ and $T_0$ for the surroundings.

*The exergy is of particular interest to the chemical engineer.*

## 2. Chemical Reactions and Free Energies

In general chemical reactions do not go to completion, and may give, say 75% conversion of reactants at equilibrium. In general we want to know where the equilibrium is.

Entropy is generally a useful term to describe what happens and how much work is needed for physical operations such as heating and mixing, but is awkward to use for chemical reactions. So another measure, related to and using the entropy was developed. This is called the **free energy** of the system, and it includes $p$, $V$, $T$ and $U$, as well as $S$ of the system. $U$ is the internal energy.

### There are actually two free energies:

- the **Gibbs free energy, (G)** is for materials in the system that start and end at the same temperature and pressure. G is particularly useful for chemical engineers and for industrial operations,

- the **Helmholtz free energy, (A)** is for materials in the system that start and end at the same temperature and volume. A is useful for the laboratory chemist who does much of his research in constant volume bombs.

# Free Energies

Values of *G* or *A* are tabulated for many substances in thermo books, or can be calculated for the substances of interest.

Now the 2nd law says that as a reaction approaches equilibrium the sum of the free energy of the reaction components, both reactants and products, lowers and **is at a minimum at equilibrium**.

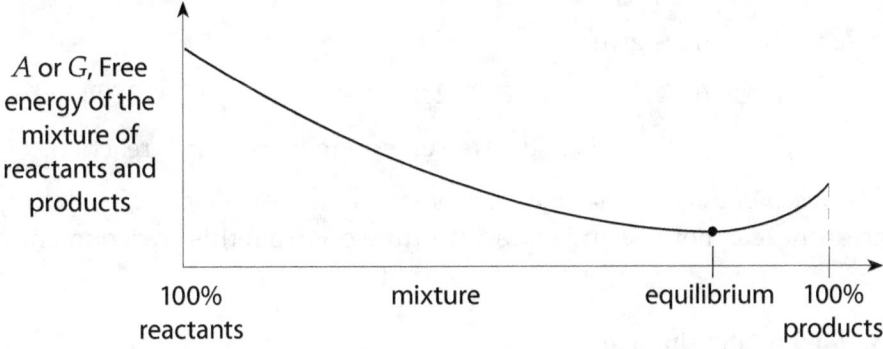

Free energies of many substances are tabulated in many chemical thermodynamic books. Here is a miniscule sample:

**Standard Free Energies of Compounds from their Elements**

| compound | $\Delta G°$, J/mol, at 298 K |
|---|---|
| elements ($O_2$, C, $H_2$, Si) | 0 |
| $CO_2$ | -394 380 |
| $H_2O$ | -237 190 |
| CO | -137 280 |
| $C_6H_{12}O_6$ (glucose) | -909 400 |
| $SiO_2$ | -800 000 |

*Let us show how to use free energies*

### Example 4. Life on Sweet Mars

Mars has an atmosphere rich in carbon dioxide, it lacks oxygen. Let us suppose that the land is sprinkled with glucose, a crazy idea.

*Could creatures survive there living off glucose and $CO_2$ as follows:*

$$C_6H_{12}O_6 + 6\ CO_2 \longrightarrow 6\ H_2O + 12\ CO \quad \ldots \quad \Delta G_{rx} = G_{final} - G_{initial}$$

**Solution.** Calculate the Gibbs free energy for this reaction

$$\Delta G_{rx} = 6\ G_{H_2O} + 12\ G_{CO} - G_{C_6H_{12}O_6} - 6\ G_{CO_2}$$

Replacing values gives

$$\Delta G_{rx} = 6\ (-237\ 190) + 12\ (-137\ 280) - [(-909\ 400) + 6\ (-394\ 380)]$$

$$= +205\ 180\ J \ldots \text{ the free energy increases with reaction.}$$

*This reaction would not advance to any significant extent by itself,* so living creatures can not live and extract useful work from this environment.

### 3. Mixing and Unmixing

Mixing creates randomness and generates entropy. Unmixing requires doing work and it reduces the entropy.

Consider the naturally occurring process of fresh river water flowing into the ocean. Here the waters mix, entropy is produced but no useful work is obtained. However the 2nd law tells that with clever engineers at work we should be able to produce useful work (while reducing entropy production), in the limit as much as $W_{max}$. If you are able to calculate it you may be surprised to find that it is the energy equivalent to that produced from water falling over a 231m high dam!

This shows that the country's rivers could supply a substantial fraction of our country's energy needs. Now here's a real challenge to our engineers of tomorrow.

Opposed to this mixing, to separate fresh water from the ocean one needs to supply, in the ideal case, the energy at least equivalent to a 231m high dam. In real situations (reverse osmosis plants in Saudi Arabia and elsewhere, or multistage evaporators) the work needed is much greater than this 231m high dam.

## 4. Entropy and Information

An increase in entropy of a system as it goes from state 1 to state 2 is directly related to the decrease in information, $I$, that we have about the system. Thus

$$\Delta S = S_2 - S_1 = -k\,(\Delta I) = -k\,(I_2 - I_1)$$

As one <u>increases</u> the other <u>decreases</u>. This is the famous relationship for which Leo Szilard received the Nobel prize. Here are some examples to illustrate this relationship.

**Example 5.** Where's the Canary?

A cage contains one canary. I wait until the canary is on the right side at which time I slip a divider into the cage.

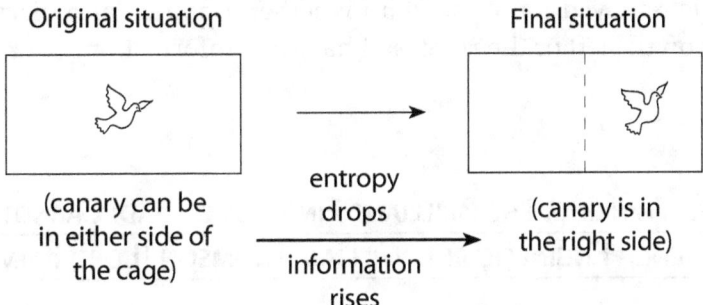

By this action calculations will show that S drops as I rises. The information about where the canary is, is in agreement with our everyday commonsense.

**Example 6.** Card Shuffling.

The sketch tells it all.

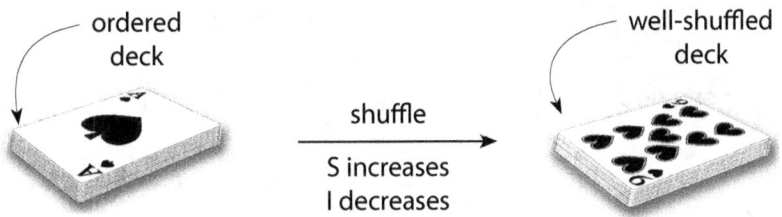

## 5. The Mixing of Gases

A container is divided in two parts, with oxygen on the left and nitrogen on the right. The divider is then removed and the gases mix.

Calculations show that the S of the tank of gases rises, and the I decreases.

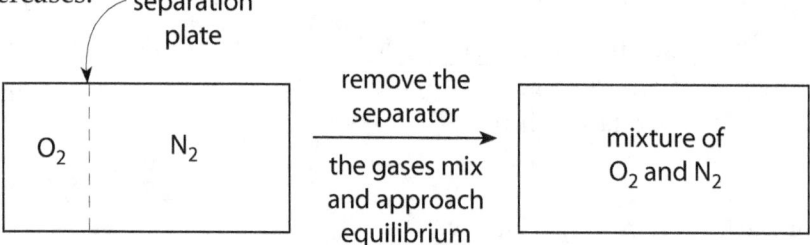

*Note:* originally I know that each oxygen molecule is on the left. After the plate is removed I am uncertain whether a particular molecule of oxygen is on the left or the right, so I have lost information.

## ADVANCED THINKING, THE BRILLIANT FINDINGS OF SADI CARNOT, A YOUNG FRENCH ENGINEER, AND WILLIAM THOMSON (LORD KELVIN)

**Finding 1**  The Reversible Heat Engine, and the fact that it is Unique

First of all, remember that experience shows that heat can only flow from high to low temperature; not the other way. Then consider a source of high temperature heat at $T_1$, and a cool sink at $T_2$. Let $Q_1$ units of heat leave $T_1$, and let it produce $W$ units of work in a heat engine and let $Q_2$ units of heat be rejected to $T_2$. This is shown below:

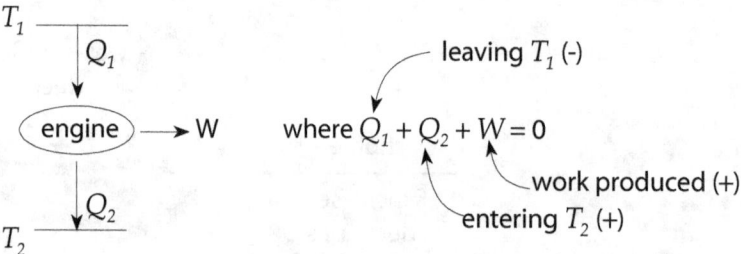

# What Sadi Did

and let a heat pump or refrigerator work the other way:

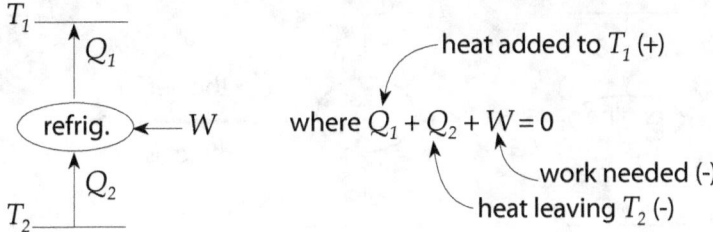

and let us call a **reversible engine** one that can work either way, as engine or pump.

We can imagine a number of heat engines working between $T_1$ and $T_2$ to produce a given amount of work, say 5 units. We show five of these below:

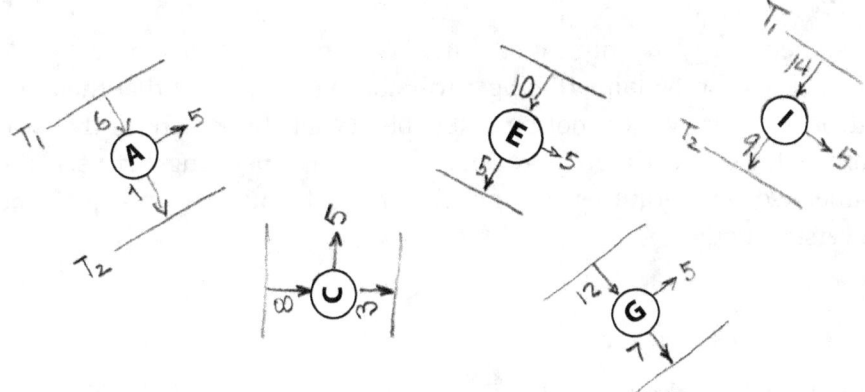

*How many of these can be reversible? maybe one? two? or maybe more than two?* We show here that only one of these engines can be reversible.

## Solution

Suppose that two of these engines are reversible, for example engine **C** and engine **E**. Then reverse engine **E** and connect it to engine **C**

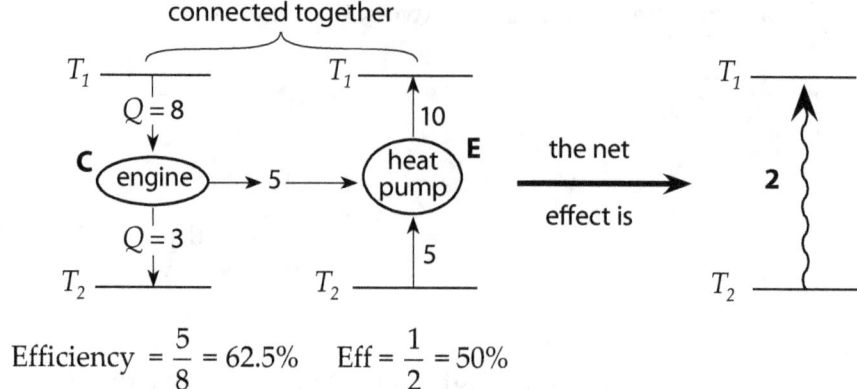

Efficiency = $\frac{5}{8}$ = 62.5%   Eff = $\frac{1}{2}$ = 50%

With this connection no work is needed nor is any produced, and the net effect is that heat flows uphill by itself. But this **violates the 2nd Law.**

So whenever you think you have two heat engines of different efficiencies, both claimed to be reversible, you will find that the more efficient of the two cannot be reversible. What this means is that you cannot have two different reversible engines operating between the same two temperatures, $T_1$ and $T_2$. At most you can have just one reversible engine.

**Finding 2**  This shows that the Reversible Engine is the Most Efficient of Heat Engines

Let us suppose that engine E is the reversible engine. Next let us show engines

**A, B, ... H,** with their thermal efficiencies $W/Q_1$ and their total entropy changes $\Delta S_{total}$. Also, let the temperatures of source be $T_1$ = 10 and of sink $T_2$ = 5.

# The Most Efficient Heat Engine

Let us line up these engines from the highest to lowest efficiency

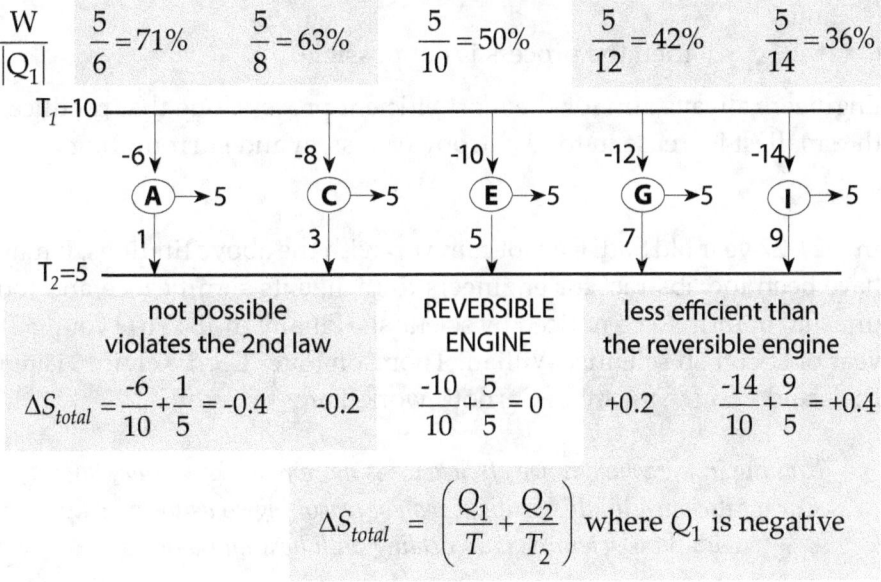

Engines **A, B, C,** to **D** are more efficient than the reversible **E**, but as shown in Finding 1 these engines violate the 2nd law. Then looking at engines **F, G, H, I** we see that they can exist but they all are less efficient than reversible **E**.

Thus the reversible engine is the most efficient of heat engines, with efficiency given by Eq 1,

$$\frac{W}{Q_1} = \frac{T_1 - T_2}{T_1}$$

**Finding 3** The most efficient of Heat Engines

Consider the heat engines of the above sketch. Their entropies are shown below the engines. So when $\Delta S_{total} = 0$ you have a reversible engine.

In general for all sorts of engines and refrigerators, as well as other non-thermo processes such as mixers, separators, electricity generating dams in rivers, chemical reactors, eating candy bars, etc., the same conclusions follow:

- If the process actually takes place then $\Delta S_{total} > 0$
- The most efficient process has a total entropy change, $\Delta S_{total} = 0$, and
- If $\Delta S_{total} < 0$ then the process is not possible.

Engineers always aim for the most efficient process, one that produces the smallest increase in total entropy of system and surroundings.

In 1824 27 year old Sadi Carnot came up with the above findings. It may have been too abstract for engineers to realize its significance and too unusual in form to be noticed by scientists. Finally in 1845 the young 21 year old Scottish scientist William Thomson (later Lord Kelvin) visited Paris and became aware of Carnot's work. Later he wrote:

> *Nothing in the whole range of Science is more remarkable than how Carnot dreamed up this brilliant way of reasoning so as to come up with these scientific laws dealing with heat and work.*

**"Thermo is a young man's game!"**

## PROBLEMS

**1. The Most Efficient of Heat Engine, The Carnot Engine**

A reversible heat engine receives heat from a heat source at 500K and rejects heat at 280K.

*What should be the heat input rate to power a 10 kW (of work) heat engine?*

**2. The Use of Dormant Volcanos**

Close to Zig-Zag, Oregon, by the side of a dormant volcano, is a large 1 km³ underground field of fractured quartz-rich rock ( $\rho = \frac{2650 kg}{m^3}$, $C_p = \frac{800 J}{kg \cdot K}$ ).

The average temperature of the rock is 500K, of surroundings is 280K.

*a) I wonder how much useful energy is stored in this rock (joules)?*

*b) How long could a giant coal fired power plant (of 1 GW output) run with this amount of energy?*

**3. Power from the Mixing of Waters.** Thermo tells that it takes at least 23 atm of pressure ($\Delta p$ = 23 atm) to squeeze fresh water through a filter from ocean water. So the reverse says that with the right device we should be able somehow to extract 23 atm of pressure from water when fresh water mixes with ocean water.

*We are trying to invent such a device, and if we are able to extract all the possible power from the Columbia river as it flows into the Pacific Ocean (7188 m³/s) how much power can we recover ($\dot{W}_{recovered} = \dot{v} \int dp$)? Is it equivalent to 1 MW, or the energy output of a giant coal-fired power plant (1 GW), or what?*

4. **Plutonium Production.** During full production of plutonium, the Hanford atomic energy facility in Washington State continuously took in and discharged about $10^5$ m³/hr of Columbia river water for cooling purposes. The river is at an average temperature of 10°C, the discharged water was at 90°C.

   *Theoretically how much power could have been recovered from this hot water stream?*

5. **The World of Science Fiction.** Science fiction writers have visualized worlds in which creatures lived on silicon compounds in place of carbon compounds. They ate sand and excreted silicon and oxygen

   $$SiO_2 \longrightarrow Si + O_2$$

   *Does this make sense from the point of view of thermo?*

CHAPTER **14**

# FIELDS - GRAVITATIONAL, ELECTRIC AND MAGNETIC

## CONCEPTS IN SCIENCE

In trying to come up with a useful picture of our world and universe, science has proposed all sorts of ideas and concepts; for example:

> ether, caloric, energy, entropy, and very recently in astronomy:
> spin networks, black holes, loop quantum gravity, spin foams,
> time advancing in ticks, time not existing between ticks.
> (*Scientific American,* Jan. 2004)

Some of these concepts were found to be useful for a while, and then were discarded, for example - ether, caloric. Others turned out to be very useful - Newtonian mechanics, Einstein relativity, etc., and still others turned out to be rationalist inventions and considered to be bunkum. As Michael Schermer wisely stated in his *Scientific American* article of Jan. 2004

> *No one talks bunkum with a better vocabulary than*
> *those who lace their bunkum with scientific jargon.*

Here is an example of a group of concepts which were created recently (see "Our Growing Breathing Galaxy" *Scientific American,* Jan. 2004, pg. 66) to be part of science, bamboozled scientists into thinking that it was, but was in fact pure bunkum. This article says, and I quote:

> *The spin networks that represents space in loop quantum*
> *gravity accommodate space-time by becoming what we call*
> *spin "foams". With addition of time the spin networks become*
> *two dimensional surfaces.*

Elsewhere the authors add time to 4-dimensional space-time to come up with a 2-dimensional surface! That just takes my breath away. And again "time advances in discrete ticks but does not advance between ticks", and so on. This is a modern wonderland of Alice's. However, as the author says:

> *Everything I have discussed is theoretical*

This is a pure rationalist dream. It is not science.

Here in this chapter we will talk about vector fields which turn out to be very useful concepts in understanding the behavior of gravity, electrostatics and magneto statics.

## Vector Fields

Consider a big tank of water, very smooth flow of water entering at a source (+) and leaving at a sink (-). Water flows everywhere in the tank, so this is called a flow field. The flow paths for the water are **streamlines** or **lines of flow** (in general called field lines), and we can associate an arrow, called a **velocity vector** to each and every point in the flow field to show the direction and magnitude of the flow. When streamlines are close together, such as near the source and near the sink, the arrows are long and flow is fast. When the streamlines are far apart flow is slow. All this is shown in the figure below.

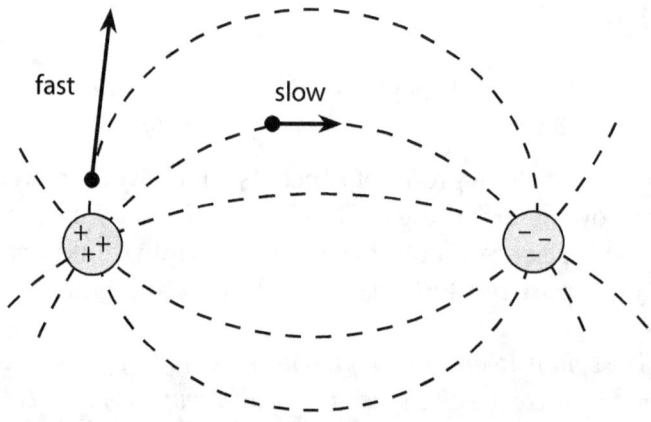

# Gravity

This method of symbolic representation is not limited to velocity fields. It can be generalized and applied to all kinds of **vector fields**.

We will discuss
- Gravitational Fields
- Magnetostatic Fields
- Electrostatic Fields

## GRAVITATIONAL FIELDS

We could (at least in our imagination) perform experiments in the vicinity of our planet by dropping little pebbles here and there, at the surface and miles out in space, measure how fast they fell and accelerated (9.81 m/s² at the earth's surface) and then map the force vectors. We would also note that the field lines all aim at a common point which is at the center of the earth.

In the field representation of water flow we saw that a separation of streamlines meant a diminution of velocity of flow. By analogy we may suspect that the divergence in our field lines here means a diminution in acceleration, or decrease in weight of a body with distance from the center of the earth. This would be easy to map with the use of a spring scale which would give you the direction of the force and its magnitude (weight):

$$W = mg \quad \text{or} \quad F = ma$$

The force field we establish is the **gravitational field** of the earth and we should call the force exerted on unit mass at any point the local field intensity.

## 14.4

**Question:** What is the local gravitational field intensity at the surface of the earth?

Newton pondered about this problem after the famous apple hit him on his more famous head and finally came up with a shrewd guess. His biggest clues were Kepler's laws concerning the planetary motion as well as the motion of the moon around the earth. Without going further into this fascinating story of how he cracked open the mystery let us give you his conclusions. Here are the generalizations that he induced:

**Every mass in the universe attracts every other mass** (1)

Now this is only qualitative. He would like a quantitative expression determining how much the attraction was. This he obtained by the so-called inverse-square law.

$$F = C / d^2, \quad [N] \qquad (2)$$

where C is some constant and $d$ is was the distance between the centers of the two masses.

He combined (1) and (2) and finally obtained what is known as **Newton's Law of Gravitation**

$$F = G\, m_1\, m_2 / d^2, \quad [N]$$

where $m_1$ and $m_2$ are the two masses involved

and G is the gravitation constant = $6.67 \times 10^{-8}$ $dyne \cdot cm^2 / gm^2$

$$= 6.67 \times 10^{-11}\ N \cdot m^2 / kg^2$$

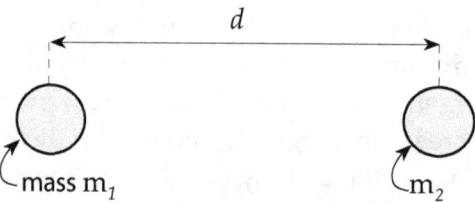

The equation we use in mechanics for Newton's Second Law at the surface of the earth ($F = ma$) is just a special case of his Law of Gravitation where $m_2$ = mass of the earth and $d$ = radius of the earth, (the distance between my belly button and the center of the earth) which are constants

$$F = G\, m_1 m_2 / d^2 = m_1 (G\, m_2 / d^2) = m_1 a = m_1 g = 9.81 \cdot m_1, \quad [N]$$

## ELECTROSTATIC FIELDS

When a substance such as glass is rubbed with a dry cloth it may mysteriously attract light objects such as bits of paper or a little feather. The feather would gently approach the glass, touch it and then a mysterious thing would happen, the feather would suddenly be repelled back into the air! It is as if an invisible wind (or force) had blown the feather onto the glass and then miraculously reversed itself to blow the feather away. Also a thin blue spark could sometimes be made to leap from the tip of a person's outstretched finger to the rubbed glass, and the experimenter may even feel a slight tingling or a shock. Moreover this shock can be accompanied by a crackling noise.

Thales (-500 BC) did similar experiments with amber, which the Greeks called elektron. William Gilbert (1600) studied this phenomenon and called it electricity, and today electricity is one of the bases of our modern societies, powering electric lights, electric motors and computers.

**These forces of attraction or repulsion are caused by electric charges.**

These are atomic in nature and are extremely tiny and thousand of millions of them together will produce forces only equal to the weight of one mosquito. But if you collect a large enough number together you can get the enormous effect of the bolt of lightning in a thunderstorm

Consider the field and the equipotential surface around a single point charge. It will look like this.

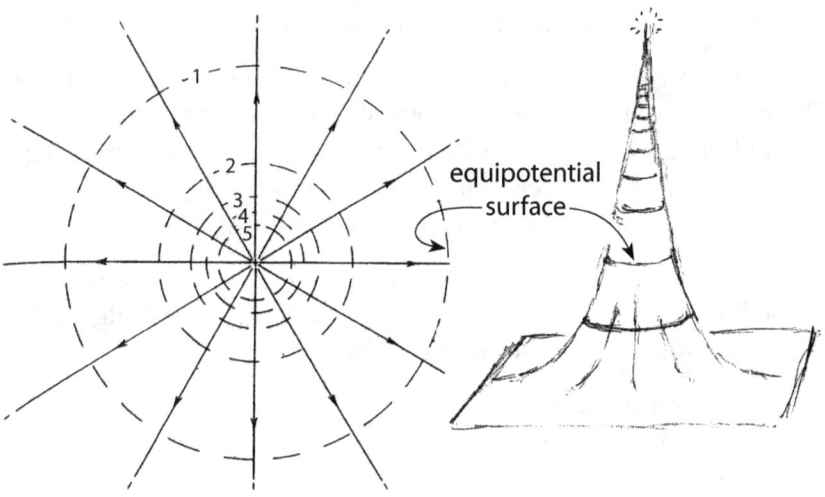

Field lines and equipotential surfaces in the region around two equal and opposite point charges, one positive, one negative, are shown:

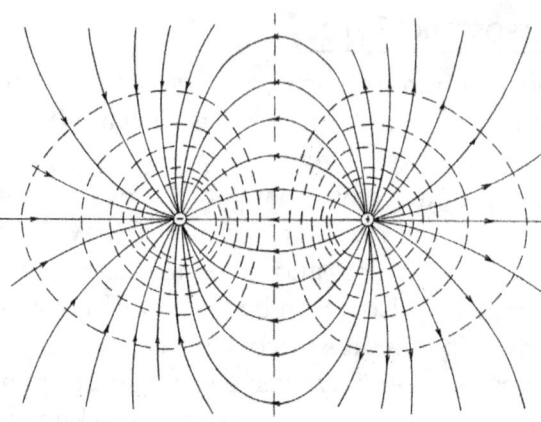

The electrical force of interaction of these two point charges depend

(1) on the magnitude of the two electric charges $q_1$ and $q_2$ located in the field
- two positive (or two negative) charges repel each other
- a positive and a negative charge attract each other

(2) on their distance apart, $d$

(3) on the intervening medium, measured by the dielectric constant, K

Coulomb devised experiments and came up with the following answer to this question of attractive force between two like charges

where:
$$F \propto \frac{-q_1 q_2}{4\pi K d^2}$$

$F$ is the force of interaction (attraction), N

$d$ = distance between the two charges, $m$

$q$ = electrical charge. In the SI system the basic unit is the coulomb, with symbol C. This is the point charge which repels an identical point charge 1 $m$ away with a force of 1 N. In the cgs system the basic unit is the smaller e.s.u. (the electrostatic unit) or the abcoulomb, where 1 coulomb = 1 C = 10 abcoulomb = 10 AbC = 10 e.s.u.

K is a constant called the kielectric constant which depends on the medium. Carendish experimented and found that it depended on the nedium around the charges. Thus

K = 1 for a vacuum
>1 for all substances
= 81 for water

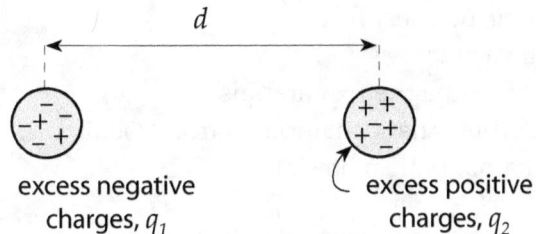

## MAGNETIC FIELDS

Ancient legend tells of a Cretan shepherd named Magnus who one day found that his feet were attracted to the earth by the nails in his sandals. He dug around into the ground to discover the cause and found a mysterious rock which was responsible for this power. This property called magnetic was known in many ancient societies to be associated with a certain kind of rock, which we call magnetite today.

The Chinese are said to be the first to suspend such rocks from a fiber in order that the earth north-south orientation that invariably took place might be useful for navigation.

Both Peregrinus and Gilbert (1600) did careful observations which lead us to today's magnets with their north and south poles, where N-N repel, S-S repel, and where N-S attract each other.

In a fashion similar to that used for charges in electricity (Coulomb's Law) we say that the force of attraction between two like magnetic poles $P_1$ and $P_2$ can be given as

where:
$$F \propto \frac{-P_1 P_2}{4 \pi \mu \, d^2}$$

$F$ = attractive force for a N-S system, and a repulsive force between a N-N or S-S system.

$P$ = magnetic charge or pole strength. In the SI system the basic unit is called the coulomb, with symbol C. It is the point charge (N pole) which repels an identical charge (another N pole) one meter away with a force of 1 N. In the cgs system of units the basic unit is called an e.m.u. (the electromotive unit) or the abcoulomb, with symbol AbC.

1 e.m.u. = 1 AbC = 10 C . . . and . . . 1 C = 0.1 AbC = 0.1 e.m.u.

μ  = magnetic permeability
   = 1 for a vacuum
   = 0.1 for paramagnetic materials
   >> for ferromagnetic materials - iron, cobalt, nickel
   <1 for diamagnetic materials

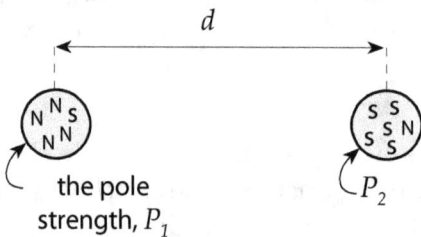

the pole strength, $P_1$      $P_2$

## PROBLEMS

1. According to Google, the mass and diameter of the earth and of the moon are

   $m_e = 6 \times 10^{24}$ kg      $m_m = 7.35 \times 10^{22}$ kg

   $d_e = 25.5 \times 10^6$ m      $d_m = 6.95 \times 10^6$ m

   In my space suit I weigh 180 lbs. *I wonder how much I'd weigh on the moon?*

2. Water issues from a point source and flows in all directions in a radial 3-d flow pattern. If the local velocity is 10 cm/s 3 m from the source, *what is the velocity 10 m from the source?*

3. What is the force of gravitational attraction between a well-fed 100 *kg* husband and his well nourished wife of 80 *kg*? Both can be considered to be approximately spherical. Assume the distance between their centers to be 1 *m*.

4. Find the acceleration of gravity due to the earth in the vicinity of the moon, which is approximtely 60 earth's radii away from the center of the earth.

## Field Problems

5. Can you determine the mass of the earth from the knowledge that 1 $kg$ mass weighs (or is attracted by the earth by) 9.8 newtons and that the radius of the earth is 6370 $km$?

6. Compute the force of interaction of two point masses, 1 $kg$ each placed 1 $m$ apart in a vacuum.

7. An electric charge of $q = -2$ C is placed between a charge of $q_1 = +4$ C and $q_2 = +10$ C so that the 3 charges lie on a straight line.

   *Where must q be placed so that it does not want to move?*

8. Yup and Yop, standing at two distant points on a clear dark night, shine their flashlights at you. The flashlights are identical. However the light from Yup's flashlight is 100 times as bright as Yop's. Yup is 1 $km$ from you.

   *How far is Yop from you?*

9. From the discussion in this chapter and the fact that the circumference of the earth is 40 000 $km$ determine the average density of the earth.

10. Suppose a rope is broken by a force of 100 kg. If you need it to lift a 64 kg body with the rope vertically what is the maximum acceleration you could give to the body?

11. a) How far would a ball drop in 2 s at the earth's surface?

    b) How far would the same ball drop on Neptune's surface?

    *Note:* Neptune has twice the mass of the earth and its diameter is twice the earth's diameter.

12. How far apart in a vacuum must two electrons be if the force of electrostatic repulsion on each electron is just equal in magnitude to the weight of the electron?

    *Data:* The electron mass is 9.11 • $10^{-31}$ kg and its charge is 4.8 • $10^{-10}$ E.S.U.

13. At what altitude above the earth's surface in miles would the acceleration of gravity be approximately 16 ft/s? Take the radius of the earth to be 4,000 miles.

    A magnetic pole of 120 units strength is 10 cm distant from a dissimilar pole of 10 units strength. Find the force of interaction between them in dynes.

14. Two magnets 5 and 15 cm long with pole strengths of 100 and 200 magnetic units are placed end to end with like poles 1 cm apart. What is the force of interaction of the magnets in dynes?

15. One gram of monatomic hydrogen contains 6.023 • $10^{23}$ atoms, each atom containing one positively charged nucleus and a single outer electron. If all the electrons in one gram of hydrogen could be concentrated at the north pole of the earth and all the nuclei at the south pole, what would be the force of attraction between them?

    *Data:* The electrical charge of an electron is 4.8 • $10^{-10}$ e.s.u.

CHAPTER **15**

# MEASURES OF THIS AND THAT

In the past we had all sorts of ways to measure things, for example:

- in England the last joint of King Henry's thumb (now displayed in Westminster Abbey) was chosen as **the inch**
- In Afghanistan the distance between the tip of your nose when you face left and your outstretched right fingers was chosen as **the yard**
- In India the area of ten cricket pitches was chosen to be **one acre**. It still is today.

There was a great diversity of units all over the world but in the industrial age they coagulated into a few systems

- the inch-pound-second (ips) system
- the centimeter-gram-second (cgs) system and
- the meter-kilogram-second (mks) system.

In 1959 the science world got together in Paris and came out with the last great change, called in French "le systeme internationale", or the **SI system**. We use this here.

In the SI set of units there are seven basic measures

| Unit | Name of Unit | Symbol |
|---|---|---|
| length | meter | $m$ |
| mass | kilogram | $kg$ |
| time | second | $s$ |
| electric current | ampere | $A$ |
| temperature | Kelvin | $K$ |
| luminous intensity | candela (added in 1960) | $cd$ |
| amount of substance | mole (added in 1972) | $mol$ |

All other derived units are defined in terms of these seven basic units, for example

| | |
|---|---|
| velocity | $v = m/s$ |
| acceleration | $a = m/s^2$ |
| pressure | $Pa = pascal = N/m^2$ |
| force | $N = newton = m \cdot a = kg \cdot m/s^2$ |
| work | $J = joule = force \cdot distance = kg \cdot m^2/s^2$ |
| power | $W = watt = J/s = VA$ |
| electrical | $V = volt = W/A = (J/s)/A$ |

## RELATION BETWEEN SI AND OTHER SYSTEMS OF UNITS

### A. Newton's Law relates force, mass and acceleration

$$F = ma = kg \cdot \frac{m}{s^2} \qquad \text{where } N = \frac{kg \cdot m}{s^2}$$

newtons, N — at earth's surface the acceleration of gravity

$$a = g = 9.806 \frac{m}{s^2}$$

### B. Length

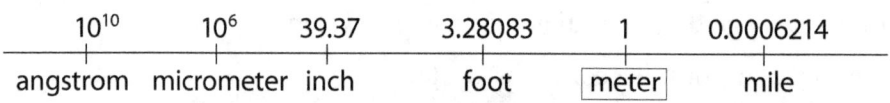

### C. Volume

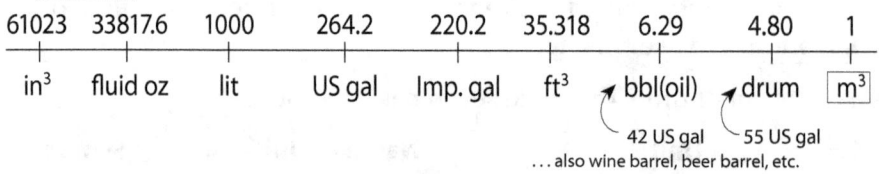

### D. Mass

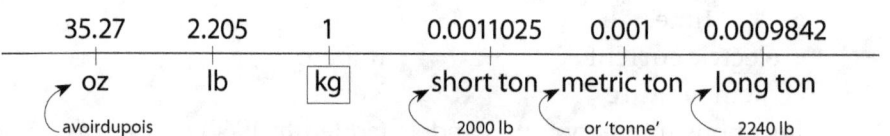

# UNITS

**E. Pressure** the pascal = $\frac{force}{area}$: $1\,Pa = 1\,\frac{N}{m^2} = 10\,\frac{dyne}{cm^2}$

$1\,atm = 760\,mm\,Hg = 14.696\,psi = 14.696\,lb_f/in^2 = 29.92\,in\,Hg$
$= 33.93\,ft\,H_2O = 101325\,Pa$

$1\,bar = 10^5\,Pa$ ... close to 1 *atm*, sometimes called a *technical atm*

$1\,inch\,H_2O = 248.86\,Pa \cong 250\,Pa$

**F. Work, Energy and Heat**

the joule = force · distance: $1J = 1\,N \cdot m = 1\,Pa \cdot m^3$

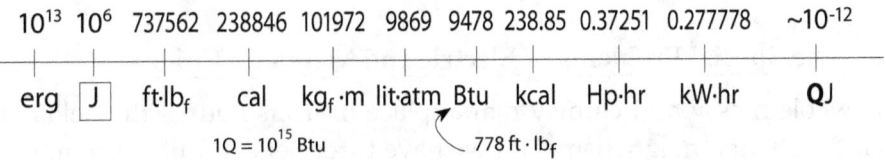

*"For those who want some proof that physicists and chemists are human, the proof is in the idiocy of all the different units which they use to measure energy."*

—Richard Feynman

**G. Power** the watt = $\frac{work}{time}$: $1W = 1\,\frac{J}{s} = 1\,\frac{N \cdot m}{s}$

```
   10⁶    1341   1341    1     10⁻³   ~10⁻¹²
────┼──────┼──────┼──────┼──────┼──────┼────
    W      Hp     kW     MW     GW     QW
```

**H. Molecular Weight** ← We should call this molecular mass

In SI units: $(mw)_{O_2} = 0.032\,\frac{kg}{mol}$

$(mw)_{air} = 0.0289\,\frac{kg}{mol}$ ... etc.

## I. Ideal Gas Law: $pV = nRT$

Gas constant: $R = 8.314 \dfrac{J}{mol \cdot K} = 1.987 \dfrac{cal}{mol \cdot K} = 8.314 \dfrac{Pa \cdot m}{mol \cdot K} =$

$0.08206 \dfrac{lit \cdot atm}{mol \cdot K} = 1.987 \dfrac{Btu}{lb \, mol \cdot K} = 0.7302 \dfrac{ft^3 \cdot atm}{lb \, mol \cdot K}$

## J. Bigger and Smaller SI prefixes

| Factor | Prefix name | Symbol | Factor | Prefix name | Symbol |
|---|---|---|---|---|---|
| $10^{12}$ | tera | T | $10^{-12}$ | pico | p |
| $10^{9}$ | giga | G | $10^{-9}$ | nano | n |
| $10^{6}$ | mega | M | $10^{-6}$ | micro | μ |
| $10^{3}$ | kilo | k | $10^{-3}$ | milli | m |

## K. The Special Problems of Electric and Magnetic Units

A whole messy and clumsy marketplace of units haunts the fields of electricity and magnetism, and we have three sets of units in common use today.

1) The **SI** (or System International). This is the standard today and is useful for large force and energy systems

2) The **cgs** (centimeter gram second). This is useful for laboratory or small systems

3) The English system. This is dying out today.

| The SI system | The cgs system |
|---|---|
| $F$ = force = N [newton] <br> = $ma = kg \cdot m / s^2$ | dyne = $cm \cdot g / s^2 = 10^{-5}$ N |
| Work = (force)(distance) <br> = $N \cdot m = J$ [joule] | 1 erg = $10^{-7}$ J |
| Power = $J/s$ = W [watt] = $kg \cdot m^2 / s^3$ | 1 erg/s = $10^{-7}$ W |
| m = mass [kg] | g [gram] = $10^{-3}$ kg |
| q = electrical charge = C [coulomb] | 1 e.s.u = 1 electrostatic unit |
| P = pole strength <br> = magnetic charge = C [coulomb] | 1 e.m.u = 1 electromagnetic unit <br> = abcoulomb = 1 AbC = 0.1 C |
| A = electric current [ampere] | |

# Chapter 16

# Cosmology

*Shouldn't we set aside magical explanations,
and instead look for scientific ones?*

Scientists, philosophers, deep thinkers and shallow thinkers have all occasionally pondered about

- how the universe started,
- how old it is today,
- how big it is today
- and will it die, and if so then how, and when?

Let's see what is the experimental evidence today from which we should make our guesses.

When man looks up at a clear moonless night sky what does he see? Stars. Yes, stars everywhere. And if asked how many, he'd likely say 'millions'. But in fact all he could see will be about a thousand.

With binoculars, with telescopes up to the giant Palomar 200 inch, or the Hubble space telescope, he'd be able to see more and more stars, even millions in all directions. Our sky seems to be splattered with stars, double stars, stars surrounded by planets, nebulae (clouds of interstellar gas or dust), groups or clouds of stars, and galaxies (collections of millions, billions, or trillions of stars held together by gravity). On the following page we show some pictures of galaxies.

## 16.2

The spiral galaxy Messier 83 with its millions and millions of stars.

Side view of a galaxy.

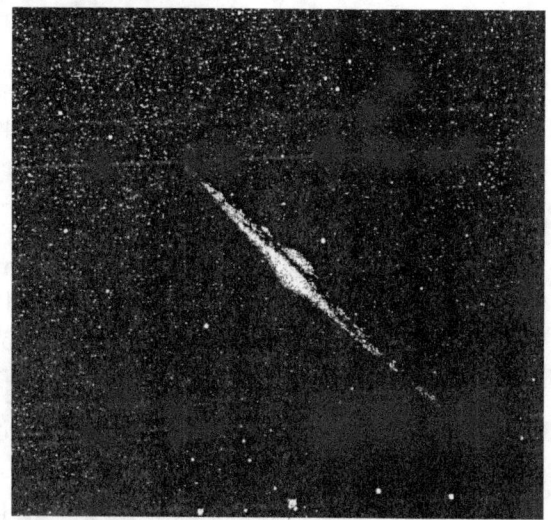

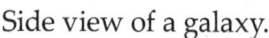

Our sun's position in the Milky Way

Actually our sun with its tiny planets is part of a giant galaxy called the Milky Way. On a clear moonless night in summer, with no bright streetlights shining or moon showing you can see the Milky Way glowing dimly across the sky.

So there are stars in the sky, everywhere we look. How far away are they, and are they moving, if so, how fast, and how old are they? Let us consider these questions.

## Einstein's Theoretical Studies

Einstein expected his General Theory of Relativity to agree with his picture of the static universe. However in trying to solve his equations Einstein found that he had to introduce a fudge factor, called the **cosmological constant** in order to successfully construct a model which agreed with his preconceived idea of a steady state universe (1917).

Russian scientists struggled with Einstein's equations. Finally one of them, A.A. Friedmann (1922) in trying to get rid of that cosmological constant in Einstein's equations, found that in his derivation Einstein had made an error. Friedmann corrected this error and came up with a surprising theoretical solution. His unexpected result was that the universe had to be non-stationary, either expanding or contracting. It could not be stationary.

At first, Einstein, whilst approving of Friedmann's mathematics, refused to accept the idea of an expanding or contracting universe. He strongly believed that the universe was immutable and forever unchanging. However, after some discussions with Friedmann, Einstein finally was won over and accepted Friedmann's result, as he reflected in his book, "The Meaning of Relativity" (Princeton NJ 1953). Later he would say that his introduction of the cosmological constant was the greatest error he'd made in his life.

## Models for the Universe

Man has come up with two types of guesses for our universe:

- **First** — The infinite, the steady state or static universe which has existed and will exist for ever and ever.
- **Second** — The finite universe which started some time ago, and eventually would have to end.

## 16.4 CHAPTER 16 • COSMOLOGY AND THE UNIVERSE

Since ancient times some societies have favored one or other of these views. Before we were able to make any experimental measurements (until the early 20th century) some societies looked at the sky as being a moving ceiling studded with fixed stars (ancient Greece, historical Chinese, Hindu, and others). As opposed to this, an example of a finite universe comes with the Jewish and Christian bibles and the Islamic Koran, which speak of a seven day creation.

In the early 20th century, man made some experimental findings which affected his ideas.

### How Far Away are the Stars?   *Miss Leavitt's Findings* [1]

When Magellan sailed around the world he noted and reported that he saw two curious clouds of stars on the southern skies. Today we call these clouds the Greater Magellanic and the Little Magellanic Clouds.

Early in the twentieth century Henrietta Leavitt was a research assistant at the Harvard College Observatory. Her job was to catalog the stars of the southern sky from photographs sent from the Harvard southern station in Arequipo, Peru. She focussed on the particular stars that got brighter and dimmer on a regular periodic basis that were found in the Little Magellanic Cloud. These are now called the **Cepheid Variables** because the first one studied by Leavitt was in the constellation Cephei.

From her catalog of two-thousand-four-hundred of these cepheids plus her detailed examination of twenty-five of these in particular Leavitt discovered an interesting fact; the brighter the variable star the longer its period , the dimmer the shorter the period.

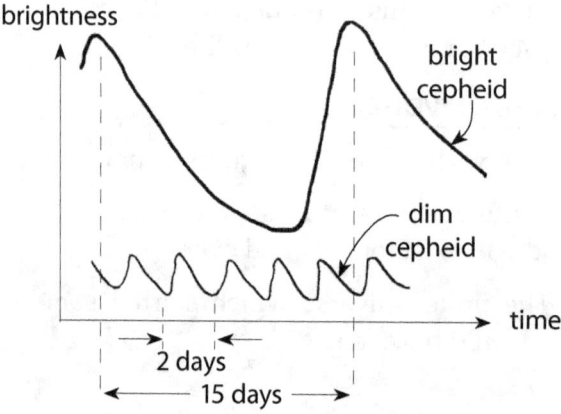

She then guessed that all the cepheids in a cloud were about the same distance away from us. So if she saw a cepheid elsewhere in the sky with the same period but which was only 1/4 as bright as her cepheid then it had to be twice as far away as her cepheid. Also if you could see a cepheid in some distant galaxy which was one millionth as bright as her standard cepheid then it had to be one thousand times as far away as her standard one.

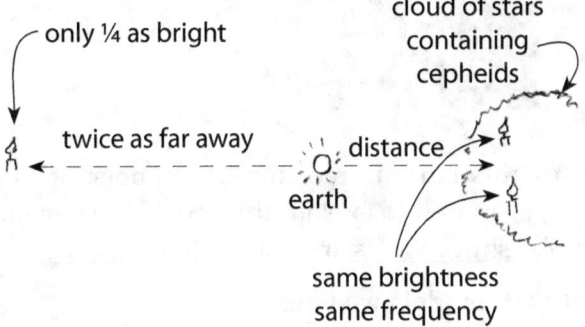

This brilliant finding, first reported by Leavitt in 1912, allows us to determine cosmic distances, even in far away galaxies. This finding should have earned her a Nobel prize, but these prizes only go to living scientists. But she died young, in 1921, before the significance of her findings became known.

## REDSHIFT OF STARLIGHT    *Hubble's Findings [2]*

The second experimental discovery comes from starlight. When you look at the spectrum of starlight you will see the colors: blue-green-yellow-orange-red, but with a careful examination you will also see some black lines in the rainbow of colors, each coming from a specific atom - hydrogen, helium, boron, etc.

By heating these materials in the laboratory we can determine the exact location of the black lines on the spectrum. These lines are then the signatures of the materials. However if we look at the starlight from a star we will see that its black lines are displaced sometimes more and sometimes less, depending on the star, but nearly always towards the red. This was Edwin Hubble's discovery, and it is called **the Redshift**.

**16.6**        CHAPTER 16 • COSMOLOGY AND THE UNIVERSE

from our laboratories
⟶

from a star or galaxy moving away from us
⟶

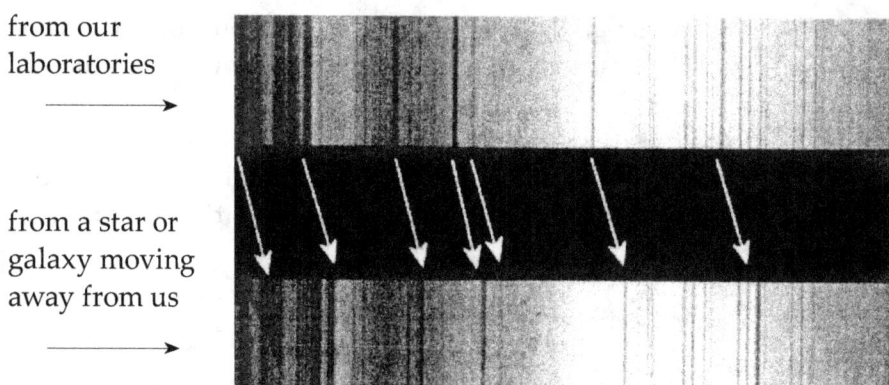

Now what causes this curious redshift? A number of guesses and models have been proposed to explain this redshift. They fall into two classes: the Doppler shift models and the friction models.

**For the Doppler shift models we have:**

   a- The Big Bang (1927) [3]
   b - The Continuous Creation (1948) [4]
   c - Alfven's Plasma (1966) [5]

**For the frictional effect models we have:**

   a - deSitter's (1971) [6]
   b - Zwicky's (1925) [7]
   c - The Tired Light (2007)

Let us look at these.

### THE DOPPLER EFFECT GUESS FOR THE REDSHIFT

**One explanation** for the Redshift says that it is caused by the Doppler effect. This is the basis for the most popular group of models for the universe today. We explain it as follows:

When a locomotive rushes towards you blowing its whistle, you will hear one tone, but as it passes and moves away the tone drops. Knowing the amount of drop, say from middle-C to G-sharp, you can tell how fast the locomotive is moving. This is called the **Doppler effect**.

By looking at the spectra of the stars we see, according to the Doppler guess, that those that are close to us are moving away slowly, those far away that can only be seen by the largest of telescopes are moving away at monstrous speeds of over 200 000 km/s, and even closer to the speed of light. Unbelievable! Today's Hubble's space telescope is looking into this curious phenomenon.

## HUBBLE'S LAW

From Leavitt's distance measures of stars, galaxies, and everything in the universe, and with Hubble's redshift measurements plus its assumed Doppler velocities we have the following picture of our universe.

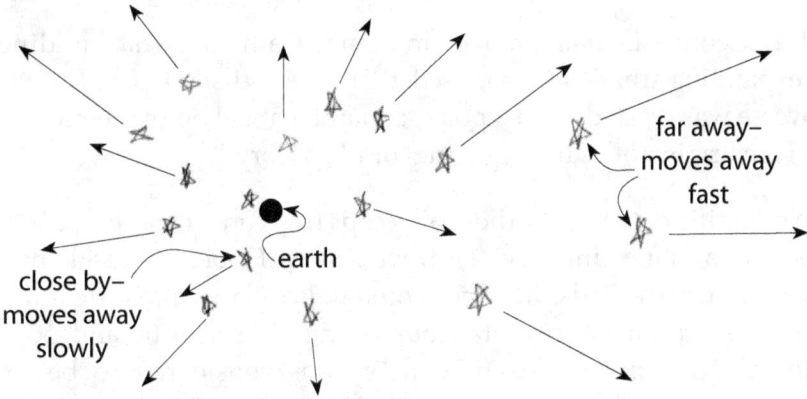

Combining Leavitt's findings with his own, Hubble ended up with what we call Hubble's Law with the Hubble constant.

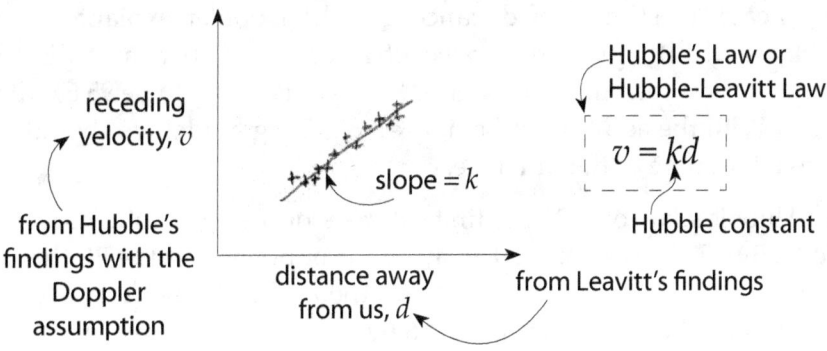

This is today's basis of experimental astronomy and Hubble is considered astronomy's top experimentalist of the 20th century.

Thus with this model nearly everything in the universe is moving away from us wherever we look, the furthest away galaxy being 13.6 Byr away and moving away from us at 95.6 % the speed of light.

## The Big Bang Model (1927)

Until early in the 20th century it was believed by just about all scientists that the stars are an eternal and practically unchanging ornament decorating the night sky, and that the universe was essentially unchanged since the beginning of time and will be so forever in the future. Then came the discoveries of Leavitt and Hubble in the early 20th century.

In 1927 Georges Lemaitre a Belgian Roman Catholic priest, building on an expanding universe, proposed that in the distant past the whole universe was created from a point, a mathematical point. Lemaitre was thus the originator of the Big Bang, or BB, theory [3].

The scientific community did not accept this conclusion because they expected a static universe. However, after Hubble's world-shaking observation of the Redshift which could indicate an expanding universe there was an almost instantaneous change of scientific and popular opinion. The universe from then on was considered to be in an expanding state, starting with the Big Bang about 15 Byr ago.

## The Age of the Universe from the BB model

Using Leavitt's estimate of distance and the Doppler explanation of Hubble's Redshift gives the speed of recession of that most distant galaxy we have seen in the universe. We find its speed to be 95.6% that of light. With the addition of Lemaitre's Big Bang model we should be able to tell the age of the universe.

From Hubble's law of 1929 his first estimate of the age of the universe gave 1.8 Byr. This was laughed away by the opponents of the BB theory because it was nowhere near Einstein's theoretical 15 Byr. It was even younger than the age of the earth, 4.6 Byr.

Around the 1940s astronomers became aware that Cepheids came in two types. So 20 years after Hubble's original age estimate, Walter Baade [9] using the newly developed Mt. Palomar 200 inch telescope recalculated the age of the universe to be 3.6 Byr.

Then Baade's student, Sandage [10], tried again and did a bit better still, 5.5 Byr. This result crept a bit closer to Einstein's value of 15 Byr.

Today we have all sorts of estimates for the age of the universe, see

| | | |
|---|---|---|
| *Einstein's Greatest Blunder,* | 2 ~ 3 Byr | (pg 99) |
| D.Goldsmith, Harvard (1995) | 10.3 ~ 15.5 Byr | (pg 97) |
| | 8 ~ 9 Byr | (pg 98) |
| | 8 ~ 13 Byr | (pg 98) |
| | 15 ~ 20 Byr | (pg 99) |
| | 14 ~ 16 Byr | (pg 110) |
| | Einstein says < 16 Byr | (pg 41) |

*The Evolutionary Universe*                           5 Byr
   G. Gamow, Scientific American ( Sept. 1976)

*Will the Universe Expand Forever?*           8 ~ 18 Byr
   J. Gott et al, Scientific American (March 1976)

*Magnificent Universe,*                            10 ~ 15 Byr
   K. Croswell, Simon and Shuster (1999)

Alfven, a sharp critic of the BB model, attacked that model as representing

> *"an antiscientific attitude and a revival of myth"*

Such arguments against the BB model were scorned by the large majority of mainline astronomers. Nevertheless they were taken seriously enough that a public debate on the subject was arranged in 1972 by the Astronomy Section of the American Association for the Advancement of Science. It was a duel between two opposing ideas and it ended in a standoff, with no agreement. The two opposing camps hurled manifestos at each other such as [5]:

- **From the anti-BB camp:** the BB approach follows theoretical prejudice and is closer to religion than to science. It had become a faith, a belief, not a science.
- **From the BB camp:** we claim that our model passes all believable experimental tests to date.

**Is the BB model a part of scientific truth? What do you think?**

### Let us try to reason here how old the universe is today from the BB model.

For this imagine that today we observe two galaxies A and B on opposite sides of us, far far away, and speeding away at 95.6% the speed of light. Since galaxies **A** and **B** were originally created by the BB at time t = 0, and then flew off in opposite directions at very high speeds, say at 95.6% the speed of light, they would need at least 13.6 / 0.956 = 14.2 Byr to get to where they were when they sent out the light which reaches us today, after their 13.6 Byr travel time *(please read this sentence carefully)*.

So the age of those galaxies **A** and **B** today, or of the universe as a whole, will have to be about 14.2 + 13.6 = 27.8 Byr, or older.

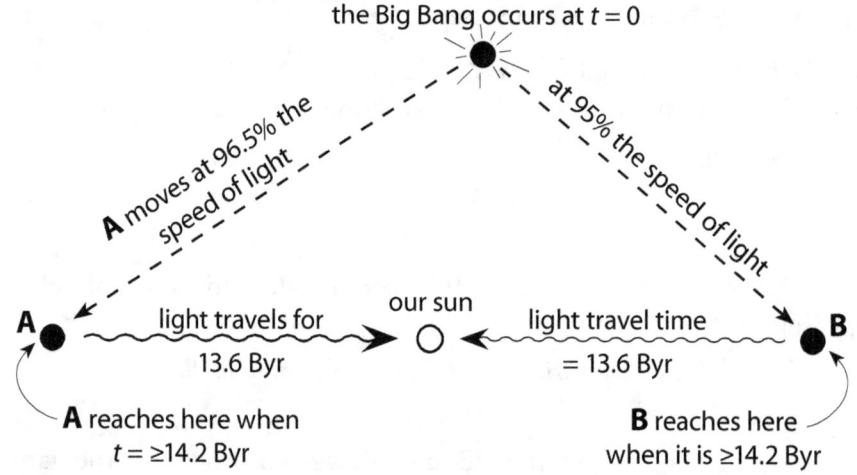

If **A** and **B** start slower than 95.6% the speed of light (today's latest idea) then it would take even longer than the 14.2 Byr to get where they were 13.6 billion years ago, so the age of the universe would be greater than 27.8 Byr.

This time is very much greater than all the estimated times reported in the literature. Also if the Big Bang occured near **A** or near **B** then the universe will have to be older than the above estimate.

So the age of the universe from the BB model is
        from past reports  2 - 18 Byr
        from Einstein's theory  15 Byr
        from the argument given here  > **27.8 Byr**

Here is a question: Are all the experimental data too small or is the above calculation about galaxies **A** and **B** wrong?

## CONTINUAL CREATION (CC) MODEL (1948)

Hoyle (of Hoyle, Gold and Bondi) et al, rejected the central idea of the BB model, that the whole universe started at a point and that it had an age of roughly 15 Byr. In fact Hoyle coined the term Big Bang to ridicule it. Hoyle's group said that the universe existed forever, and accepting the Doppler explanation for the Redshift, they say that the universe is definitely expanding. The drop in density of material in the universe could only be countered by creating and introducing matter atom by atom (one atom per century per km cubed) in the dark empty spaces between the stars to just compensate for the material lost by expansion.

But why did the density of the universe have to remain constant?

## QUESTION ABOUT THE CC MODEL

How does it explain the creation of stuff from nothing? This counters one of the very basic laws of science, the first law of thermodynamics, or the conservation law.

Cosmologists came up with an most ingenious answer. At the creation point of the positive particle, is also created a matching "negative particle". How simple. The laws of science are now not violated, and scientists have no basis for complaining—very tricky.

In addition, these negative particles, of course, exert negative gravitational effects. This makes them repel each other and this could be the reason why the universe is expanding instead of contracting, as it should with ordinary matter. How brilliant.

Why have we not seen this negative universe? They say because we have not tried to use our negative telescopes.

Those thinkers talk of black holes and white holes, also negative mass, negative pressure, black stars and so on and so on. To me this is just a fantasy and a fairytale. It is not science, and I can't understand how scientists can be bamboozled by these myths.

### ALFVEN'S "PLASMA UNIVERSE" MODEL (1966)

Alfven [5] then proposed his own model, that the universe originally consisted of a thin mixture of stuff. With the action of gravity and with time, maybe billions or trillions of years, we don't have any idea how long, little coagulations formed, then larger ones which collected more and more stuff, and finally the stuff condensed into a monstrous and probably hot blob. This became unstable, exploded giving the present universe. In this model no original creation is assumed. We show this graphically:

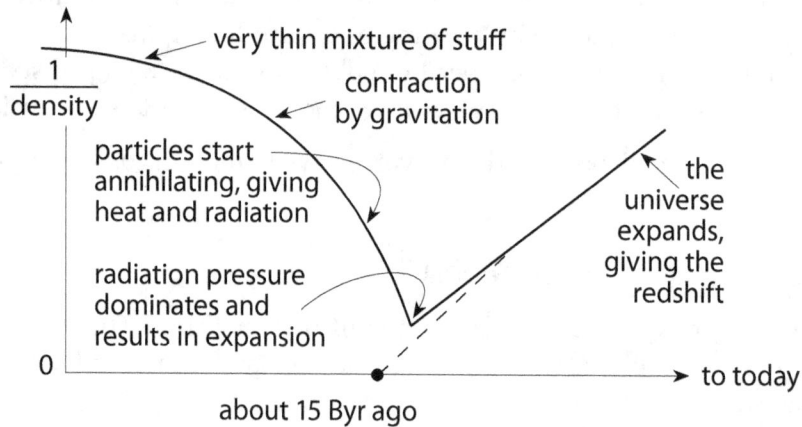

### THE GRAVITATIONAL REDSHIFT MODEL

According to Einstein's General Relativity theory we learn that when a light beam leaves a strong gravitational field its energy is lowered by the pull of the gravitational field and it wants to slow down. But relativity does not allow it to slow down, so what is affected by the gravitational pull is the frequency of the light and its associated wave length. The frequency decreases and correspondingly its wave length increases. This is called the Gravitational Redshift, and it was mentioned by Einstein in 1917.

On the other hand, when a light beam enters a strong gravitational field it will experience the opposite effect, a Blueshift. So for a beam that enters and then leaves, in effect passes through a gravitational field, these two effects cancel each other and the light will not show any shift.

However, when we analyse the spectrum of the starlight from our night sky we see definite Redshifts everywhere. The gravitational Blue-and-Redshifts certainly are not their cause. We must look elsewhere.

## THE FRICTION EFFECT MODELS

The Doppler effect models disagree and contradict experience and with the basic ideas of science (particularly thermodynamics). Let us look for an alternative explanation for the Redshift.

## THE TIRED LIGHT MODEL (TL) OF THE UNIVERSE

This model assumes that the Redshift is caused by some sort of frictional or gravitational drag on the light wave as it travels for its millions or billions of light years in to and out of gravitational fields of stars and galaxies to eventually reach us. According to Einstein's General Theory of Relativity the frictional drag cannot slow light so the only way it can loose some of its energy is to vibrate more slowly, thus giving the Redshift, whether the light is going into or out of a gravitational field.

What happens to this lost energy? Does it end up as the rather recently discovered microwave radiation which seems to come from all directions in space? It is a low level "hiss" which can only be detected by the most sophisticated microwave equipment, and its temperature is about 2.7 K. Maybe.

We have not been able to verify this guess in the laboratory because the speeds are astronomical, but in principle it could be verified because it would be an actual measurement, not some sort of magical quantity which defies the basic laws of science.

In this model we do not imagine that all the matter of the universe started as squashed into something smaller than the last joint of your little finger, nor do we propose that matter is created out of empty space, popping out as little kernels of popcorn once every 100 years in each $km^3$ of space. Neither do we propose that the light sources, stars and galaxies, are flying about at unimaginable speeds, close to that of light, and other such magical events.

## 16.14           Chapter 16 • Cosmology and the Universe

Now the farther the light has to travel the greater is the color shift (more ins and outs of gravitational fields) and of course we can always match it with an observed Hubble Redshift. With this explanation we end up with a stable universe, without stars and galaxies wandering about at close to the speed of light.

### Other Early Friction Effect and Related Models

In 1917 William de Sitter's theoretical model [6], somewhat like Einstein's, predicted that clocks would appear to run more slowly the further away they were away from the observer, meaning that light would be more redshifted the larger the distance between source and observer. From his mathematics the lines in the spectra of distant stars and galaxies must therefore be displaced towards the red, thus the redshift.

In the 1920's Fritz Zwicky [7] proposed his **Tired Light Model**. It was somewhat similar to this TL model, and it still has a few followers today.

But for far away galaxies coming towards us, as is Andromeda, how can this model explain the observed Blueshift?

### How to Weigh Scientific Models

In general, in weighing scientific models Richard Feynman [5] stated

> *First you guess. Don't laugh; this is the most important step.*
> *Then you compute the consequences.*
> *Compare the consequences with experience.*
> *If it disagrees with experience your guess is wrong.*
> *In that simple statement is the key to science.*
> *It does not matter how beautiful your guess is or how smart you are,*
> *if it disagrees with experience, then it is wrong.*
> *That's all there is to it.*

### Myths Beyond Science

> *"Sometimes I've believed as many as six impossible*
> *things before breakfast."* —Alice

In science all statements form a consistent whole. So you can go from one statement to another; from Newton's laws to Einstein's relativity without contradiction. However, in theoretical astronomy, which explains our observations and speculates about the past and the future, most of today's popular models start with a group of statements <u>which contradict science</u>.

Unfortunately we have not been able to design any experiment to check these extra-science comological 'truths' against scientific facts, however we do know that they are inconsistent with the fundamental ideas of science. In fact they are delusional. So when we combine these 'truths' with scientific statements we end up with a set of hybrid statements that seem like science but are not. And they may look convincing. but as Hubble warned:

*"Not until the empirical explanations are exhausted need we pass on to the dreamy realms of speculation"*

Here is an extra-science statement widely accepted in cosmology:

*The whole universe, stars, galaxies and everything else that we can and cannot see, in fact everything, started as a point.*

What in the world does that mean? It makes no sense.

By accepting the explanation that the Hubble redshift was caused by the Doppler effect, by ignoring all other possible explanations we are led to all sorts of magical phenomena. In particular , it led to the Big Bang model, which says that the universe started as a point. But this goes against the truths of science, in particular:

*you cannot create matter from nothing*

From this magical assumption cosmologists have created stories which violate the truths of science, in particular the laws of thermodynamics.

In Chapter 2 where we talk about truth and knowledge we point out that most men believe and accept at least two systems of truths, one from religion or myth, the other from science.

Here we point out that even in the field of astronomy we find an offshoot of science, for example the Big Bang, which we accept as true but is inconsistent with the truths of science. Other models such as Hoyle's Steady State Universe also rest on the assumption of creation from nothing, so these models also are beyond science.

*For a smile, here is a little story to end this chapter.*

A touring professor started his favorite lecture with "Sixty-five million years ago the earth was trod by dinosaurs".

He was immediately interrupted by a well-meaning old lady: "You mean sixty-five million and eight years ago, don't you?"

"Why do you say that?"

"Because I heard you give this same lecture eight years ago", explained the old lady.

The professor was shaken but he bravely continued on to cosmology and said that according to today's most favored theory, the Big Bang, the universe experienced its giant explosion at its creation, about 15 billion years ago. The stars would shine and after about 15 billion more years they would burn out. That is the time when the universe would die a cold death. When he stopped and asked for questions the same old lady asked:

"Sir, did you say that the universe will die 15 million or 15 billion years in the future?"

"15 billion, madam."

"Thank you, sir. I am so relieved."

## Problem

**The Lost Inca Treasure.** In the catacombs under the engineering building of the Universidad Central del Ecuador careless students accidentally punched a hole through the wall and came across an enormous mountain of one inch square pieces of gold, worth about 1.5 gazillion dollars. Why 1 in squares? That was puzzling. Professor Jorge Medina and his faculty put their minds to this problem and came up with a brilliant guess. They thought that this could well have been the treasure hidden by Inca General Ruminahui from Francisco Pizarro, the maurauding 1531 AD Spanish invader.

These 1 inch squares were the key to the mystery, and it is explained as follows. This gold was produced by an ancient and secret Inca process where you melt a small bath of pure gold, pour it into a square pan, and let cool to make an 8 x 8 inch cake. Then carefully and accurately cut the cake into 4 pieces (see Fig 1).

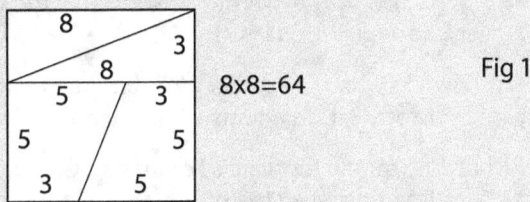

Fig 1

Rearrange the four pieces to make a 5 x 13 inch cake (see Fig 2)

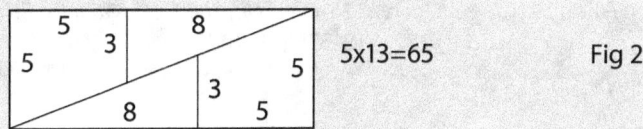

Fig 2

You now have gained, in fact created, an extra square inch of gold So cut off 1 square inch from the cake, save it, and then melt the rest and repeat the process to get another piece of gold; and so on, and on, and on.

This was the technology that helped make the Incas society dominant in South America until Francisco Pizarro and his rapacious band came and destroyed the Inca civilization.

## REFERENCES

1. Henrietta Leavitt, *Periods of 25 Variable Stars in the Small Magellanic Cloud*, Harvard College Observatory Circular,173 (1912).

2. Edwin Hubble, *A Relation between Distance and Radial Velocity among Extra-Galactic Nebulae*, Proceedings of the National Academy of Science, **15**, Number 3, (March 15, 1929).

3. Georges LeMaitre, *Annals of the Scientific Society of Brussels*, **47A** 41 (1927).

    Simon Singh, *The Big Bang, the Origin of the Universe*, Harper Perennial (2005).

4. Fred Hoyle, Thomas Gold and Hermann Bondi, (1948); also see Helge Kragh, *Cosmology and Controversy*, Princeton U. P. 172-179 (1966).

5. Hannes Alfven, *Worlds- Antiworlds: Antimatter in Cosmology*, San Francisco, W. H. Freeman (1966).

6. Willem de Sitter, *On Einstein's Theory of Gravitation and its Astronomical Consequences*, Monthly Notices of the Royal Astronomical Society, **78**, 3 (1917).

7. Fritz Zwicky, *On the Redshift of Spectral Lines through Interstellar Space*, Proc. Nat. Academy of Science **15** 178 (1929).

8. Richard Feynman, *Feynman Lectures on Gravitation*, B. Hatfield, editor, Reading MA (1995).

9. Walter Baade, *Extragalactic Nebulae*. Report to IAU Commision 28, Transactions of the International Union of Astronomy, **8** 397-399 (1952).

10. Allan Sandage, *Current Problems in the Extragalactic Scale*, Astrophysical Journal, **127** 513-526 (1958).

CHAPTER **17**

# THOUGHTS ABOUT THE UNIVERSE

In looking at the sky on a clear moonless night with binoculars, or telescope we see

- planets
- stars
- clumps or clusters of stars
- galaxies
- black spots or bare holes in the sky
- binary stars circling each other
- nebulae or clouds of fuzzy objects, not points of light
- Cepheid variables which brighten and dim periodically
- the Milky Way
- a rare bright flash in the sky which slowly dims and disappears

No matter how powerful the telescope, this is what we see. From this we try to tell what is happening, what has happened, and what will happen.

From Chinese mythology we have:

Ancient Hindu mythology saw the universe as being created, destroyed and then re-created in a very complex manner in eternal and long cycles. At any time it was viewed as being held up by elephants.

Aristarchus of Samos (310-230 BCE) a Greek astronomer had a very different view. He placed the Sun at the center of the known universe, with Earth revolving around it, and the other planets in their correct order of distance around the Sun. He also believed that all the other stars in the sky were much, much further away since they exhibited no parallax.

Unfortunately his idea was scoffed at by ancient Greece (~300 BC onward) in favor of the theories of Aristotle and Ptolemy which viewed our world as sketched below

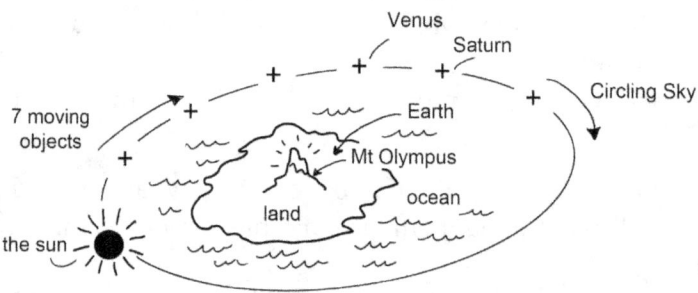

Christian Europe, until about 1500 AD, adopted these Earth-centered ideas, not Aristarchus' Sun-centered idea. In the 16th century a Polish cleric and astronomer, Nicolai Copernicus (1473-1543), assumed that if he adopted a stationary central Sun it would simplify all his calculations. This idea was revolutionary but was ignored for about 100 years.

Then came Galileo. He carefully examined the sky with the help of one of the first telescopes made, and said "I don't just see 7 moving objects in the sky, as our Christian truth claims, because I see at least 9, two of which circulate Jupiter!"

With the new instruments being developed from the 15th century onward, plus new ideas, findings and analyses by Kepler, Newton, Darwin and many, many others (including European religions) and too many others to mention or consider here, we have until the beginning of the 20th century viewed our heavens as sprinkled by things which did not move, but just stayed where they were.

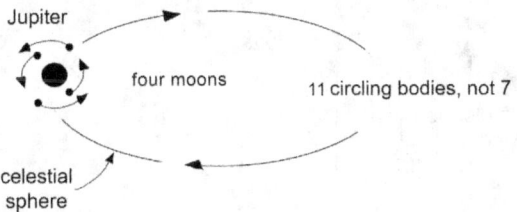

Then in 1905 an enormous revolution in our thoughts occurred followed by two experimental findings somewhat later. First of all, Albert Einstein, an assistant 3rd class in the Swiss Patent Office, just contemplating things in his spare time, which he had a lot of, dreamed up and published his Special Theory of Relativity. In addition he published three other ground breaking papers, one of which received the Nobel Prize; all this in the same year - amazing.

On the experimental side, in 1907 a young lady named Henrietta Leavitt, was hired as a Research Assistant by the Harvard College Observatory, at 30 cents/hr. Her job was to analyze the findings being sent in from Harvard's laboratory, which was located at Acapulco, Peru. She was sent 299 photographic plates of stars taken by 13 telescopes, and in 1913 published her findings. Leavitt focused her efforts on the variable stars in the Magellanic Cloud, a dim patch of light in the Southern skies, which was first noticed by Magellan (1520) when he circumnavigated the Earth. Through a telescope she noticed that the cloud consisted of thousands and thousands of stars including many variable stars. Some pulsed in a regular fashion, some slowly, others rapidly. Studying these she found that their periods of pulsation varied from about 1 day to over 50 days, but more importantly, that the brighter the star, the longer was its period. In fact she found that the logarithm of its pulsation period was proportional to its brightness, or

$$\log (\text{period}) = k (\text{brightness}) \ldots \ldots \ldots (1)$$

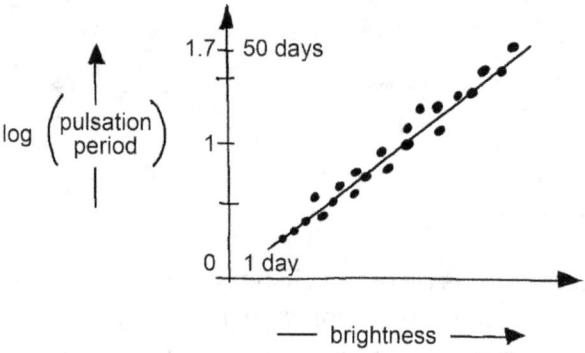

Also since the data fell into a rather narrow band, it meant that the stars in the band were all about the same distance away. So the Magellanic cloud was really a neighboring clump of stars.

Leavitt looking elsewhere in the skies found other variable stars, and if one of these had a period equal to one in the cloud, she concluded that it had to be as bright. But if it looked dimmer then it had to be further away. So if it was only 1/4 as bright it had to be twice as far away, if 1/100 as bright then it was 10 times as far away as the one in the cloud. This was a way to measure distances in the sky, and it was a **spectacular** discovery.

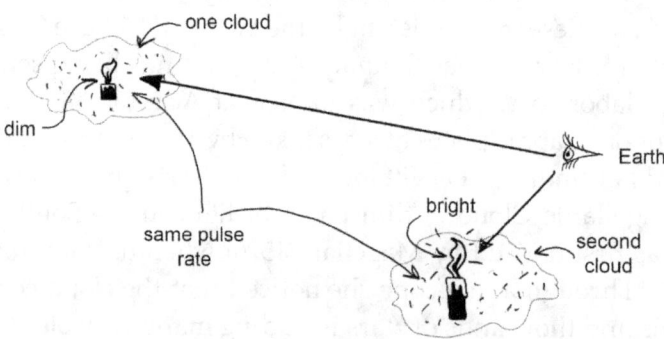

The second important finding was by Erwin Hubble, who carefully examined the spectrum of the light of stars and galaxies from his telescope images. He found that by far the majority of the images showed red shifts. Assuming that the Doppler effect was the reason for the shift, he concluded that a small shift meant a small velocity away from us, while a larger shift meant a larger velocity away from us.

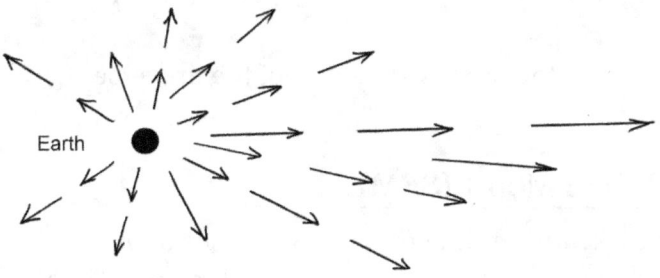

Hubble looked at his and Leavitt's results and concluded that in whatever direction examined, stars and galaxies:

- close to us in general were moving away from us SLOWLY,
- further away from us, FASTER, and
- very far away from us, as could only been seen by the most powerful telescopes, CLOSE TO THE SPEED OF LIGHT.

This conclusion meant that objects in the universe are going away from us, or else that the whole universe itself was expanding! This was a startling conclusion and it has dominated the research since then. It means that the universe is not static as was then assumed. It is MOVING, it is ALIVE! But how did it start? Would it continue expanding or would it stop and maybe (horrors) reverse itself? What had happened in the past and what will happen in the future? Such questions still intrigue scientists today.

These questions challenged cosmologists' creativity and imagination. Guesses were suggested, hypotheses forwarded, theories written about, mythical objects created, and so on. Cosmology became a wild pursuit, unhindered by the laws and rules of science, creating invisible worlds of giant suns, or with masses of miniscule dust which penetrated everything, every atom of our bodies, and so on.

To help answer these questions the Hubble Space Telescope was built and sent into orbit. It is a giant instrument with a 100 inch mirror. It cost billions of dollars and its primary purpose was to create an accurate 3-D map of the universe - to find distances between stars and to find how far away stars are from us. For rather close stellar objects, these findings use the parallax distance method. For further objects they rested on the Cepheid and Red Shift methods developed by Leavitt and Hubble.

Here are some of the proposed models of the universe.

## THE BIG BANG MODEL (BBM)

Thinking backward in time from the velocities of objects fleeing away from us today, a Belgian Jesuit priest Le Maitre, in 1925, calculated that the universe must have started about 14 - 15 Billion years ago, a date which is consistent with Einstein's General Theory of Relativity, developed just a few years earlier.

Le Maitre's theory assumed that the whole universe – millions of stars, galaxies, our sun, the Earth - all started at a mathematical point. It became unstable, exploded and flew apart, then slowed down ever since. This is called the BIG BANG Theory by Hoyle. This is a catchy term and a recent survey showed that over 70% of cosmologists today believe that this theory represents what really happened.

But an unexpected phenomenon was observed, something that did not make sense. For about the first 7 Byr after the BB the universe expanded while continually slowing as expected from the law of gravity.

**But then it reversed itself, and started to expand - faster and faster. This behavior contradicted the universal law of gravitation as we know it today.**

Does this mean that the BB model should be discarded? Or could it be that the law of gravity just didn't apply for such systems. Researchers supporting the BB model had a simple and elegant solution. At the stroke of the pen they proposed that the universal law of gravity just reversed itself. Simple. They said that somehow two very large bodies do not attract one another, and Newton's Law just does not apply here.

This brings us to the question of consistency of a system of rules. Let us take, as an example, arithmetic. We all know the rules. But let me now introduce one tiny contradiction.

    Let me assume and accept that +1 = +2

    Then using accepted math,

        Squaring both sides gives 1 = 4

        Subtracting 3 from both sides gives -2 = 1

These last two equalities above are not part of the arithmetic that we know. Similarly <u>I can prove anything that I want if you allow me to start with just one contradiction.</u>

The same type of argument applies to cosmology. Let us look at the BB Model, today's most popular . It starts by believing and accepting that the universe - galaxies, stars, our sun, the earth – all started at zero time, all squashed into one little bit of stuff, something smaller than my fingernail. Does this make sense?

By accepting the explanation that the Hubble redshift was caused by the Doppler effect, we are led to all sorts of magical phenomena. In particular to the Big Bang model, which says that the universe started as a point. But this goes against the truths of science, in particular:

> *you cannot create matter from nothing*

After the Big Bang's unexplainable explosion and expansion, this model leads us to other scientific impossibilities, that the original little dot of nothing was unbelievably hot, at millions of degrees, but in exploding and expanding it cooled to its present temperatures. This again violates the laws of science, which say:

> *A gas expanding into a vacuum does no work (because it does not push back the surroundings) hence it does not cool.*

From these two magical assumptions cosmologists have created theoretical stories which violates the truths of science, in particular the laws of thermodynamics.

Also, reversing gravity and having bodies repel each other is counter to Newton's long accepted Law of Universal Attraction. But that did not bother the disciples of the BB model. Model builders make assumptions and when the consequences disagree with fact they change their assumptions, again and again if needed. It does not bother them.

## The CONTINUOUS-CREATION MODEL (CCM)

The next most popularly accepted model is the Continuous Creation Model of Hoyle and co-workers [1945]. They scoff at the BB assumption of original creation of a fingernail of immense stuff. Their view was that the universe was originally empty. Then, sort of magically, single atoms of stuff were created, like popcorn kernels, roughly one atom every one hundred years in each one km cubed of empty space. These single bits of stuff occasionally combined and grew, then grew again, eventually creating our universe. But how was this matter created from nothing? We have no idea or explanation, and it doesn't bother them.

Continuous Creation Model

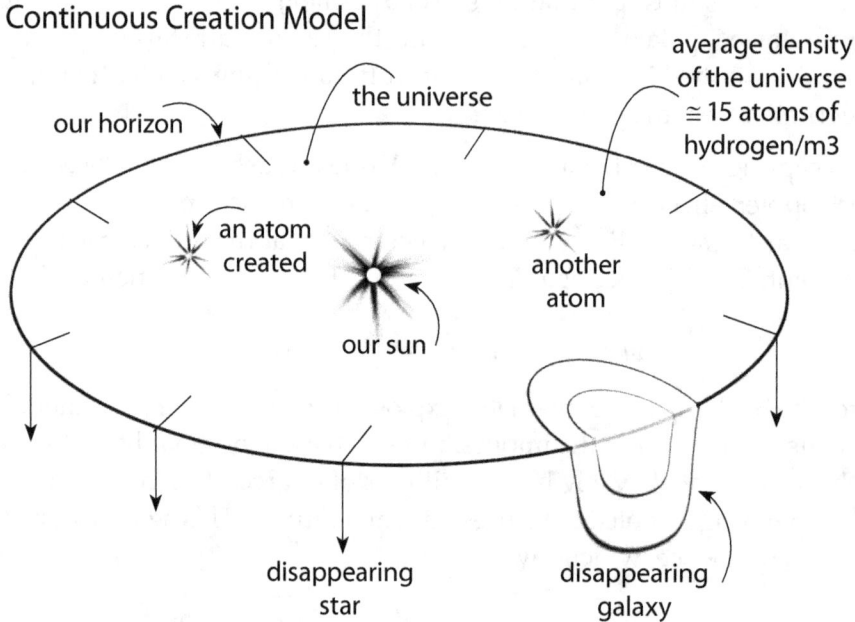

## The BOUNCING-PLASMA MODEL (BPM)

Here Alfven (1966) proposed that all the matter in the universe originally consisted of a lean mixture of stuff. Then with the action of gravity and time, maybe billions or trillions of years, we have no idea how long, little coagulations formed, followed by larger ones which collected more stuff. Finally this all condensed into a monstrous and probably hot blob. This blob became unstable, exploded about 15 Byr ago and flew apart while slowing down, to give us our present universe.

## BLACK HOLES, WIMPs, MACHOs AND HIGGS BOSONs

One of the puzzles about our universe is:

Why do observations show that the sky is very barren of stars and galaxies, while theories say that there should be many, many times more stuff in the sky; some say as much as 400 times more than is actually observed? Theoreticians answer with the words "Black Holes" which is a nice catchy term which means stuff that can not be seen, because it is invisible. How simple!

There are supposed to be two types of Black Holes:

**MACHOs** (Massive Compact Halo Objects)   and

**WIMPs** (Weakly Interacting Massive Particles).

MACHOs assume that throughout time matter wandered about and occasionally stuck together by their gravity to form clumps. These grew, and occasionally, with time, they became so massive, thousands of times the size of our sun. Their gravity became so enormous that they began to absorb, and in a sense vacuum up all the closely passing light. Hence the area of closely surrounding sky looked black. Unfortunately these black holes were only able to account for a small part of the missing mass.

WIMPs are another explanation for the missing mass. These are supposed to be extremely tiny bits of stuff which ancient Greeks called atoms, each **much, much smaller** than any known elementary particle, such as an electron, proton or neutron. They are so small that they pass through matter and through light beams without disturbing them. Most amazing! Thus millions and millions of WIMPs are supposed to pass through us and everything, everywhere in the universe every second, like a mist or fog, unobserved and undetected. Is this science? The U. of Chicago, Princeton U., AT&T, NSF, U.C. Berkeley, and others have been busy searching for this elusive stuff, supported by big money, all without success, according to C. M. Miller in Chris Miller's Home Page,1995.

## HIGGS BOSONs

Recently I read in the newspapers that scientists claim to be close to finding a sort of WIMP called the Higgs Boson, which has been called the "GOD" particle. With such a term we await with breathless anticipation for the latest news of its finding.

The Higgs Boson is a hypothetical something which is predicted to exist according to the Standard Model of Particle Physics. Leon Lederman, former head of the Batavia IL-based Fermi Laboratory, has written a book on this mysterious thing. In addition, the world's largest Hydron Collider (LHC) was built expressly to hunt for this magical stuff by the European Center for Nuclear Research (CERN). It used a 17 mile long underground circular tunnel near Geneva and it has been intensely searching for this mysterious unseen stuff for many years.

It cost billions of US dollars to build this giant laboratory and scientists claim that they are tantalizingly close to have spotted the Higgs Boson. Despite enormous efforts at both the CERN and FERMI laboratories the Higgs Boson has not yet been identified as of December 2011. But who knows, maybe tomorrow?

But in July 2012, the leader of the 2000 engineers searching for the Higgs Boson at CERN reported that of the billions and billions of collisions of the protons traveling close to the speed of light, a few dozen generated 'significant' trash which may have hinted strongly at possible evidence of bosons.

## DARK MATTER

In the late 1960's two US astrophysicists, Davis and Bahcall, decided to study the capture of solar neutrinos emitted by nuclear fusion in the sun while eliminating any other type of interfering emission, so they chose to run their experiment 1,478 meters underground in the old Homestake gold mine in Lead, North Dakota. Why there? To eliminate any cosmic ray debris from coming down and fouling the results.

After 3 months of testing with their extremely sensitive equipment the experimenters expected to have recorded about 1600 collisions of the elusive solar neutrinos with the Xenon surface of their argon isotope target. They found no collisions.

Similar experiments followed in Japan, in the former Soviet Union, in Italy, and in Canada, all showing questionable results. What did they find? Nothing, or ... ZERO.

These models introduced us to terms such as dark matter, electron neutrino, muon neutrino and so on. On the contrary, objectors feel that this kind of cosmology differs from the rest of physics by violating the basic rules of scientific reasoning. Also according to Alfven and the BB and the CC models, and all the models that rely on them are also anti-scientific and are a revival of myth.

## DARK ENERGY

Unfortunately inventing the existence of dark matter could only account for just a small portion of the calculated hidden matter in the universe. So another material, also unseen so far, had to be invented to account for the rest. Well, very recently Saul Perlmutter of U.C. Berkeley received the Nobel Prize in Physics for claiming to have made progress in answering this question. He proposed that this unknown material consisted of a mysterious DARK ENERGY, not dark matter, and that it represented about 75% of the total mass of the universe. The presence of this dark energy is supposed to lead to an acceleration of the expansion, not a deceleration of the expansion of the universe, a most surprising prediction which counters all the truths of physics as we know them today.

Today, researchers from the US and Europe are actively trying to verify the existence of this DARK ENERGY. A research station in Antarctica, a new space telescope with a 2 m diameter mirror, which will be much more powerful then today's Hubble, and a more sensitive spectral measuring device - are all in the works to help answer this question, according to Perlmutter.

By combining Leavitt's and Hubble's findings on the Doppler Red Shift and Cepheid variables Perlmutter is supposed to have made progress in solving the problem of the hidden dark energy. This got him the Nobel Prize in 2011.

It is hoped that this will soon solve the mystery of DARK ENERGY.

## The FRICTIONAL DRAG Model (FDM)

At this point let us consider a model which bypasses mystery and magic, called the Frictional Drag Model (FDM) of our universe. In our experience EVERY transformation of energy from one form to another involves friction which leads to some kind of inefficiency. This may be large, such as when operating a steam locomotive, or it may be miniscule as when using an iPod. Operations in our world ALWAYS generate some sort of friction. However, as opposed to everything else that we know, we claim that we can transport information by light with complete 100 % efficiency.

In our laboratories with the most sophisticated equipment ever used we can not observe any frictional effect or diminution of light intensity. As opposed to this, looking at light from stars we see a Red Shift, but we say that it is caused by the source of light moving away from us, the Doppler effect. For the most distant galaxies recorded by this most powerful Hubble space telescope, with its giant 100 inch mirror and using 100 hr photographic exposures of galaxies, shows an enormous Red Shift. According to the BB model this means that the galaxies are fleeing from us at more than an incredible 95% of the speed of light!

Although some of the Red Shift may be due to the Doppler effect, I strongly suspect that some portion of it could be, and probably is caused by some sort of friction of the light as it travels its millions of millions of light years across the universe into and out of countless galaxies to reach us. For a beam of light traveling that far, the accumulated friction could maybe become visible. The very, very low micro soft energy radiation recently observed coming from everywhere in the skies could be the waste from this friction rather than trash from the original gigantic cosmic Big Bang which was supposed to have started the universe at zero time. This should be tested, if possible.

## The TIRED LIGHT Model (TLM)

As an extension of the Frictional Drag Model, Ashmore's "NEW Tired Light Model" says that all, yes ALL, of the red shift is caused by a sort of frictional drag, none by the Doppler velocity effect on the incoming light beam (see Lyndon Ashmore, *"NEW Tired Light Model"* November 2003).

First of all experimental evidence of light from the stars and galaxies shows that the amount of red shift increases the further away the light source is from us, thus, twice as far away means about double the amount of red shift, or, about twice the increase in wavelength of the incoming light beam from the expected wavelength.

According to Ashmore the mechanism for this action is that the light photon in its thousands or millions of light years of travel to reach us occasionally collides or interacts with an electron, loses a tiny, tiny fraction of its energy by the collision, and then continues on its way. This lowers the energy or frequency of the light beam and this results in a red shift.

He then presents his idea of how this is done.

## COMPARISON OF THE TWO CLASSES OF MODELS OF THE UNIVERSE

The many creative extensions of the Doppler redshift models lead us to all sorts of amazing scenarios such as; an expanding universe starting at a point, or an exploding space with galaxies flying apart at faster than the speed of light, reverse gravitation, invisible dark energy and dark matter, black holes here and there, and even multiple universes, some say eleven of them, others say many more.

The FD and the TL models are unexciting and rely on a single simple unitary explanation of what happens with light. Similar to all other actions which we experience in life these all involve some kind of friction, whether large or small. This is summarized by the second law of thermodynamics, and the Tired Light model assumes that light is no exception.

We prefer the FD and TL models to the many previously proposed models involving mass creation: at a point (the popular Big Bang); multiple bangs of different sizes created here and there; single particles created continuously throughout the universe like popcorn popping; unobserved massive masses many times the size of our sun floating about; the MACHOs; and the very dense but invisible storms of incredibly small invisible particles called WIMPs invading the whole universe including you and me, and elsewhere.

Maybe we should remember Hubble's

*"Not until all empirical explanations are exhausted  
need we pass on to the dreamy realms of speculation"*

## SUMMARY and CONCLUSION

**Let us review these six Velocity Effect Models,** *see references on page 16.18*

The three main models that build on this velocity guess are the **Big Bang Model (BB)** of Georges Lemaitre (1927) [3], the **Continuous Creation model (CC)** of Fred Hoyle and his coworkers (1945) [4], and Hannes Alfvens' **Bouncing (Plasma) model** (1966) [5].

**Model 1 - the BB model** assumes that in the past, about 14-15 Byr ago, the universe all started at a point, exploded and flew apart.

**Model 2 - the CC model** assumes the universe has been expanding for ever and ever, but to keep its matter density constant it creates atoms everywhere, one by one like little popcorn kernels, roughly one atom each 100 years in each cubic kilometer of space.

**Model 3 - the Bouncing (Plasma) model** also assumes that the Doppler shift is the cause of the red shift, but also insists that only known physical laws should enter cosmology. Hence no creation of matter, either all at once (the BB model), or atom by atom (the CC model), or by any other non-physical means should enter cosmology.

This model says that long ago all the matter in the universe started to come together by gravitational attraction. About 15 Byr ago it became so dense and unstable that it experienced an enormous explosion and then flew or bounced apart, first expanding while slowing. Then incredibly, after about 7 Byr it started to accelerate its expansion rate. There is no scientific explanation for this. This violates the laws of science.

In addition, these three Doppler effect models also violate other basic laws of science. In particular, how could objects 13 Blyr to the left and right of us have both been formed at the same point 2 Byr earlier?

**Model 4 - The Gravitational effect.** According to Einstein's Theory of General Relativity we learn that when a light beam leaves a strong gravitational field, its energy is lowered by the pull of the gravitational field, and it wants to slow down. But relativity does not allow it to slow down, so what is affected by the gravitational pull is the frequency of the light and its associated wave length. The frequency decreases and correspondingly its wave length increases. This is called the Gravitational Red Shift and it was mentioned by Einstein in 1917 [6, 7].

**Models 5 and 6** [8, 9, 10] say that friction and also maybe the action of gravity on the light as it travels through the universe causes the red shift.

The latest findings of the Hubble space telescope with its newly upgraded optics shows that the universe is populated by millions of millions of galaxies, as far as Hubble is able to see. The universe seems to be endless and all this talk about its beginnings and its magical non-scientific properties are wild speculations. They may be fun to imagine and dream up, but **they have nothing to do with science**. We prefer the simple scientifically based Frictional Drag or Tired Light models.

CHAPTER **18**

# THE LAST WORD

Mankind has dreamed up a whole turbulent world of concepts, thoughts and myths. If we think about it we would see that these ideas coagulate and develop into self-consistent clumps or groups of opinions and ideas. This leads to peoples having different beliefs and ideas of what is true. This is the situation everywhere.

These groups grow and shrink with time, are absorbed by other more aggressive groups (think of the missionary salesman), and they combine and split apart (Roman Catholic and Protestant churches). They are like living organisms (single-celled and multi celled creatures, up to man). It reminds us of nature at work.

One of these coagulations started about 500 years ago and grew to become science as we know it today. Some of the imaginary concepts, yes imaginary, that underpin science are: force, energy, entropy and fields. Other concepts were part of this scheme but then were dropped from this network: ether, life force, phlogistons and epicycles.

Since ideas such as truth appear in various forms, but have different meaning in these coagulations, this can cause much confusion. However the concepts accepted in science have special restrictions attached to them—they must be consistent with each other, and they must also agree with what we see in the world around us. Those are their special features. This little volume focuses on science and its sidekick, technology.

P.S. A very small part of this book was written on the first of April.

# INDEX

2x2,
    non zero-sum game 9.6-8
    zero-sum game 9.4
    zero-sum game, example 9.5

## A

Abundant numbers 4.2
Acapulco 17.3
Acceleration 10.6
    of expansion 17.10
Adam Smith 9.11
Age, from the BB model 16.9
    of earth 2.2
    of the universe 16.8
Alfven's universe 16.11, 17.8
Alternative hypotheses 7.3
Amicable number pair 4.2
Analytic statements 2.4, 4.3
Applied mathematics 4.5
Aquinas, Thomas 1.4
Argument,
    complex 3.3
    necessary, probable 3.2
    simple, to make 3.1
    valid, cogent 3.2
Aristarchus 1.2, 17.2
Aristotle 17.2
Arrow of time 13.5
Ashmore, Lyndon 17.11
Authoritarian approach to
    knowledge 12.1
Availability 13.11
Avogadro's Law 11.6
Axioms 4.3

## B

Baade, Walter 16.8, 16.17
BB model 16.7, 17.6
Bayes Theorem 6.3 - 6.5
Bede, Venerable 2.1, 6.2
Bend, an Oregon invention 13.3
Best line through data 7.15
Big Bang model 16.7, 17.6
Black Holes 17.8
Bombers and fighter 9.17
Bondi, Hermann 16.17
Boole, George 3.4
Boson 17.9
Boyle's Law 11.4
BP model 17.8
Branch of mathematics 4.3
Bruno, Giordano 1.4
Btu 12.5
Bulgarian Game, example 9.17
Bunkum in science 11.8

## C

Cal, or the calorie 12.5
Calorie, big 12.5
Canary 13.15
Candela 15.1
Card shuffling 13.15
Carnot, Sadi 13.16
    engine, example 13.21
    heat pump 13.5
Carroll, Lewis 4.1
CC Model 16.10
Central limit theorem 6.6, 7.5
Cepheid Variable 16.3
CERN Laboratories 17.9
Charles' Law 11.4, 11.6
Checkers 4.3
Chess problems 4.13
Chicken 9.14
Chinese checkers 4.13
Chinese mythology 17.1
Church and Galileo 11.1
Coefficient of performance 13.5
Cogency 5.14
Collision,
    elastic, inelastic 10.9
Comparison of models 17.13

Conditional Probability  6.4
Concepts invented in science  14.1
Conditional Probability  6.5
Confidence,
    coefficient  6.20
    interval  7.12
COP  13.5
Continuous Creation model  17.8
Copernicus  1.4, 17.2
Correlation coefficient  7.5
Cosmic background radiation  16.12
Cosmology  Chapters 16 and 17
Coulomb  14.6
Counting  4.1
Creation,
    of matter  16.10
    of the earth  1.9
Creationist view  2.2
Criterion  9.2
Cryptograms  11.11
Cuban Missile Crisis  9.14

## D

Dark Energy  17.11
Dark Matter  17.10
Darwin, Charles  1.9, 17.3
    and tuba  7.7
De Sitter Model  16.13
Decision and game theory  9.20
Decision theory  8.1
Deduction  3.2
Deficient numbers  4.2
Desirability  8.2
Diamagnetic material  14.7
Dielectric constant, K  14.6
Direct deduction  4.4
Doppler Effect  16.6
Dorment volcanoes  13.21
Dragon  1.1
Durant, Will  1.8,
Dynamics  10.7

## E

Earth, age of  2.2
    creationist vs scientific  2.2
Einstein  6.27, 12.2, 12.8, 16.3
    comments on Galileo's  11.1
    on rationalism  1.9
    special theory  1.10, 17.3
Electric unit  15.4
Electrostatic field  14.5
Elements  4.3
Empiricism  1.7, 12.1
    birth of  12.4
Energy  12.2
Engine,
    James Watt  13.2
    maximum extractable work  13.4
    Newcomen  13.2
Entropy  13.7
    and information  13.15
    change  13.7
    work generated  13.7
    examples and applications  13.8-10
Equilibrium state  13.6
Erg  12.4
Esthetics and art  2.3
Estimating,
    a population mean  6.12, 6.21
Ether  1.9, 12.6, 17.1
Ethics and religion  2.3
Euclidean Geometry  4.7
Exergy  13.11
    example  13.13

## F

Fahrenheit temperature scale  1.6
Falling bodies  10.3-5
False statements  3.2
Faraday  2.1, 12.9
FD model  17.11, 17.13
FERMI laboratories  17.9
Ferromagnetic material  14.7
Feynman, Richard  1.11, 12.4, 16.14
Field  14.2
Finite and infinite universe  9.10
First Law of Thermodynamics  12.2
Force of interaction  14.6
Free energy,
    Gibbs  13.12
    Helmholtz  13.12
Frictional Drag model  17.12

## G

Galaxy  16.1
Galileo  12.9
    The two new sciences  1.7
    The universe  17.2
    dialogues  1.4
    and the Church  11.1
    and the Pope  1.10
    falling bodies  10.3
    First Law  10.7
    Second Law  10.7
Game,
    and decision theory  9.20
    matrix  9.2
    payoff  9.3
    with 2+ players, example  9.9
Games  9.1
Gardner, Martin  8.8
Garfield, President James  7.25
Gaussian distribution  7.3
Gay-Lussac Law  11.4
General Theory of Relativity  16.8
Generalization  5.1-3
Gibbs, W.  12.2, 12.9
Gilbert, William,  14.5, 14.7
Girvin, H. P.  1.8
Gleneden village party  9.12
Gold, Thomas  16.17
God's Particle  17.9
Gossett  7.24
Graham's Law  11.6
Grand opera  9.6-8
Gravitational field  14.3
Gravitational redshift model  16.12
Gravity  10.7
    reversal  17.7
Great Library of Alexandria  1.2
Guiness Brewery  7.11

## H

Heat engine,
    Carnot  13.4
    reversible  13.4
Helmholtz  12.2, 12.9
Hendrick van Loon  2.2
Heresy  1.4
Hershel, Ronald  6.4, 6.6
Higgs Boson  17.10
Hindu mythology  17.2
Horsepower, the HP  12.5
Hoyle, Fred  16.10, 17.8
Hubble, Edwin  16.6, 17.5, 17.13
    constant  16.7
    Law  16.7
    space telescope  17.6, 17.13
Human intolerance  2.1
Hydro Collider  17.9
Hyperbolic geometry  4.7
Hypothesis  5.3, 6.14

## I

Ideal Gas Law, consequences  11.4
Impulse of colliding objects  10.8
Incas  16.17
Independent Probability  6.4
Index of forbidden books  1.5
Induction  3.2
Inductive arguments  5.1
Inference, to make  3.1
Infinite and finite universe  9.10
Information and entropy  13.15
Informative statements  4.3
    analytic, synthetic  2.4
Intellectual  11.9
Intersection  6.3
Invented concepts  14.1

## J

James Watt engine  13.2
Joule  12.1-4, 12.9
    the unit of force  12.2, 12.4

## K

Kepler  17.3
Kelvin  13.4, 13.5, 15.1,
Kinematics  10.4
Kinetic Theory of Gases  11.5
King Bumbledorf  8.7, 8.8

## L

Laughing Dragon  1.1
Leaning Tower of Pisa  10.3
Least squares line  7.5

through data  7.3
Leavitt, Henrietta  16.3, 17.3
Lederman, Leon  17.9
Legends about our world  2.1
LeMaitre, Georges  16.7, 17.6
Leonardo da Vinci  10.2
LHC  17.9
Life energy  12.2, 12.9
Life force  12.2, 12.9, 17.1
Light, speed of  12.4
Lobachevskian Geometry  4.7
Logic  3.1
   and mathematics  4.6
Lost Inca Treasure  16.16
Loop quantum gravity  14.1
Lord Kelvin  13.16, 13.20
LS line (least square line)  7.4
   example  7.4

## M

MACHO  17.9
Magellan  16.3, 17.3
Magellanic Cloud  16.3, 17.3
Magnet problems  14.8-11
Magnetic,
   charge  14.7
   permeability  14.7
   units  15.4
Magnetostatic field  14.7
Magnus  14.7
Mathematical,
   expectation  8.4
   induction  4.4
   system  4.3
Mathematics and logic  4.6
Maximin  9.3
Mean  6.11, 6.20
Median  6.11
Medina, Jorge  16.16
Method of Trees  6.4, 6.6, 6.7
Meyer  12.1
Michelson  12.7
Micro soft radiation  17.11
Medieval art  11.9
Milky Way  16.2-3
Million Dollar Game  9.14

Minimax  9.3
Mixing  13.14
   of gases  13.16
Mode  6.11
Model,
   big bang  16.7
   continuous creation  16.10
   de Sitter  16.11
   plasma  16.12
   steady state  16.11
   the BB  16.6
   tired light  16.13
Models, evaluation of  16.14
Momentum  10.8
   conservation  10.9
Morganstern  9.9
Morley  12.7
Most efficient heat engine  13.18
Moving continents  1.10
Mt. Olympus  1.2
Music and plant growth  7.20
Myths  2.1, 16.16

## N

Natural selection  11.7
Nebula  16.1
Needham  11.10
Negative particle  16.11
Negative universe  16.11
New branch of mathematics  4.8
New TL model  17.11
Newcomen Engine  13.2
Newton  6.27, 17.3
   First Law  10.7
   Second Law  10.7
   Third Law  10.8
Newton, unit of work  13.1, 15.1-4
Newton's Law of Gravitation  14.4
Non zero-sum  9.20
   game  9.2
Normal distribution  6.12
Null hypothesis  7.3
Numbers,
   abundant  4.2
   amicable pair  4.2
   deficient  4.2
   perfect  4.2

# Index

## O
Observation 6.11
One-tailed test 6.13
Operation Research 9.18
Optimist 8.1
Optimistic rule 8.3
Outcome 6.2

## P
Pa, the pascal 15.2
Paramagnetic material 14.7
Parameter 6.11
Pasteur, Louis 11.10
Payoffs 9.3
Peregrinus 14.5, 14.7
Perfect numbers 4.2
Period of pulsation 17.4
Perlmutter, Saul 17.10
Pessimist 8.1
Pessimistic rule 8.4
Pisa, Leaning Tower of 10.3
Plasma universe 16.12
Players,
    more than one person games 9.1
Plutonium example 13.22
Pointing fingers 9.18
Population 6.11
Postulate,
    sets 4.7
    mathematical 4.3
Power 15.2
    from mixing 13.21
    the watt 10.5
Prime numbers 4.2
Prisoner's dilemma 9.13
Probability 6.1 - 6.4
    conditional 6.4
    independent 6.3
Ptolemy 17.2
Pure mathematics 4.5
Pure strategy 9.4
Pussy's lunch 9.18

## Q
Q, the queue 12.5

## R
Rationalism 1.2, 12.1
Rationalist dream 14.2
Reason 3.1
Redshift 16.5
Reductio ad absurdum 4.4
Revelation 17.1
Reversible heat engine 13.16
Riemannian Geometry 4.7
Rumanian coin flip 9.5
Ruminahui, General 16.17
Russell, Bertrand 4.5

## S
Saddle point, search for 9.3
Sadi Carnot 13.3, 13.16
Sample
    random 6.2
    size 6.11
    space 6.1, 6.4
Sandage, Allan 16.8
Santbach 1.3
Scalinger 2.1
Scatter about the LS line 7.3
Scarpia 9.6-8
Schermer, Michael 11.8
Science, basis 1.11
Science fiction example 13.22
Scientific,
    law 5.2, 11.3
    method 11.2
    theories 5.2, 11.5
    truth 1.11, 17.1
Second Law - importance
    in engineering 13.7
Shubic dollar auction 9.12
Shuffling a deck of cards 13.14
SI system 15.3
Significance level 6.15, 7.3
Slavery, Thomas 13.1
Space 10.5
Special Theory of Relativity 12.8
Spectrum of light 16.5
Spherical geometry 4.7
Spin networks 14.1
Spontaneous generation of life 11.9

SS,
    sum of squares 7.4
    model 16.11
    remaining 7.5
    removed by the LS line 7.5
    total 7.4
Standard deviation 6.12
Statistical tests 6.14
Statistics 6.10
    and mathematics 7.10
    birth of 7.10
Steam pumping engine 13.2
Stone - paper - scissors 9.19
Strategies 9.2
Student t-test 7.24
Syllogism 3.3
Synthetic statements 2.4

## T

Tired Light model 17.12
t-test,
    example 6.16
    for slope 7.5
    of hypothesis 7.7, 7.11
    one-tailed and two-tailed 7.10
t-table of values 6.17, 7.10
Taiwanese checkers 4.14
The Pope and Galileo 1.10
Theorems 4.3
Theory of evolution 11.7
Theory of numbers 4.2
Therm 12.5
Thermometer 1.6
Thomas Slavery 13.1
Thomson, Benjamin 12.1
Time 10.5
Tired light model 16.12, 17.12
TL model of the universe 16.12
Tosca 9.6-8
True statements 3.2
Truths, different 2.1
T-table 6.17
Tuba 7.20
Two-tailed test 6.13 ~ 6.19

## U

U-test of hypothesis 6.15

U. S. President Johnson's dilemma 9.13
Union 6.3, 6.4
Units 15.1-4
Universe Chapters 16 and 17
    endless 17.13
    expanding 17.5, 17.10
    thoughts about 17.1
    tired light 16.12
    continual creation 16.10
    static 16.8
    age of 2.2
Unmixing 13.14
Ussher, James 1.8, 2.1

## V

Value of a game 9.5
Value of game theory 9.11
Variance 6.12
Various mathematical problems 4.11-14
Venn diagram 3.4, 6.4
Vis viva 12.9
Vital force 12.2, 12.9
von Brunn 1.10
von Helmholtz 11.10
von Neumann 9.9

## W

Wegener, Alfred 1.10 - 11
Whewell, Professor 1.8
Wild West Game 9.18
WIMPS 17.9
Work 15.1-4
Work, the newton, N 13.4
World systems 11.1

## Y

Young, Thomas 12.2

## Z

Zero-sum, 9.20
    game 9.2
Zeus 1.1
Zwicky, Fritz 15.14

# ABOUT THE AUTHOR

Octave Levenspiel is Emeritus Professor of Chemical Engineering at Oregon State University, with primary interests in the design of chemical reactors. He was born in Shanghai, China in 1926, where he attended a German grade school, an English high school and a French Jesuit university.* He started out wanting to study astronomy, but that was not in the stars, and he somehow found himself in chemical engineering. He studied at U. C. Berkeley and at Oregon State University where he received his Ph.D. in 1952.

His pioneering book, "Chemical Reaction Engineering" was the very first in the field, has numerous foreign editions and has been translated into 13 foreign languages. Some of his other books are, "The Chemical Reactor Omnibook", "Fluidization Engineering" (with co-author D. Kunii), "Engineering Flow and Heat Exchange", "Understanding Engineering Thermo" "Tracer Technology" and "Rambling Through Science and Technology".

He has received major awards from A.I.Ch.E. and A.S.E.E., three honorary doctorates: from Nancy, France; from Belgrade, Serbia; from the Colorado School of Mines; and he has been elected to the National Academy of Engineering. Of his numerous writings and research papers, two have been selected as Citation Classics by the Institute of Scientific Information. But what pleases him most is being called the "Doctor Seuss" of chemical engineering.

*Interested in reading about Octave's life? His daughter wrote a biography of him and it is available at **Lulu.com**, just look up <u>Levenspiel.</u>

www.ingramcontent.com/pod-product-compliance
Lightning Source LLC
Chambersburg PA
CBHW081235180526
45171CB00005B/433